Tarantulas and Scorpions in Captivity

by Russ Gurley

Professional Breeders Series

ISBN 978-0-9713197-9-0

Copies available from:

ECO Herpetological Publishing & Distribution
915 Seymour
Lansing, MI 48906 USA
telephone: 517.487.5595   fax: 517.371.2709
email: ecoorders@hotmail.com
website: http://www.reptileshirts.com

T-Rex Products, Inc.
http://www.t-rexproducts.com

LIVING ART publishing
http://www.livingartpublishing.com

Zoo Book Sales
P. O. Box 305
Lanesboro, MN 55900 USA
507.467.8733
http://www.zoobooksales.com

Design and layout by Russ Gurley.
Cover design by Rafael Porrata.
Printed in China

Front Cover: The beautiful South American Tarantula, *Avicularia versicolor*. Photo by Bill Korinek of the Theraphosid Breeding Project.
Back Cover: *Opistophthalmus wahlbergii*, the Tricolor Scorpion. Photo by Russ Gurley.

ACKNOWLEDGEMENTS

I would like to thank Bruce Effenheim and Bill Korinek of the Theraphosid Breeding Project for sharing their collection. Without their passion and support, this book would not have been possible.

Bill Korinek of the Theraphosid Breeding Project happily provided photos of his beloved tarantulas and tree spiders. Jan Ove Rein of The Scorpion Files also provided some excellent photos and information. At the last minute, Bill Love of Blue Chameleon Ventures, Eric Ythier of Scorpion Fauna, and Dan Read provided photos that enhanced this book.

Theraphosid Breeding Project - http://www.theraphosidbreedingproject.com
The Scorpion Files - http://www.ub.ntnu.no/scorpion-files/index.htm
Scorpion Fauna - http://www.scorpionfauna.com
Blue Chameleon Ventures - http://bluechameleon.org

Thanks to Bob Ashley of ECO and Eric Thiss of Zoo Book Sales for their encouragement during this and other projects. They continue to set the natural history and pet book market ablaze with their innovative approach to publishing and distribution of the most informative and exciting titles on the market today.

Lastly, thanks to my wife, Fionnuala, and daughter, Cait. Their inspiration and support make it all worthwhile.

TABLE OF CONTENTS

INTRODUCTION 6

Chapter ONE: Tarantulas as Pets . . . 7

Chapter TWO: Housing 14

Chapter THREE: Feeding 21

Chapter FOUR: Breeding 24

Chapter FIVE: Caring for Spiderlings . . 37

SPECIES ACCOUNTS

Brazilian White-banded Tarantula *Acanthoscurria geniculata* . **42**
Bloodleg Tarantula *Aphonopelma bicoloratum* . **43**
Pink-toed Tarantula *Avicularia avicularia* . . **44**
Martinique Pink-toed Tarantula *Avicularia versicolor* **46**
Curly-haired Tarantula *Brachypelma albopilosum* . **48**
Mexican Red-legged Tarantula *Brachypelma emilia* . **49**
Mexican Red-kneed Tarantula *Brachypelma smithi* . **50**
Greenbottle Blue Tarantula *Chromatopelma cyaneopubescens* . **52**
King Baboon Spider *Citharischius crawshayi*. . **53**
Rose-haired Tarantula *Grammostola cala* . . **55**
Colombian Giant Tarantula *Megaphobema robustum* . **57**
Blue Bloom Tarantula *Pamphobeteus nigricolor* . **58**
Indian Ornamental Tarantula *Poecilotheria regalis* . **60**
Sun Tiger Tarantula *Psalmopoeus irminia* . **62**
Golden Starburst Baboon Spider *Pterinochilus murinus* **63**
Goliath Bird-eating Spider *Theraphosa blondi* . **65**

SUGGESTED READING 67

Tarantulas and Scorpions

SCORPIONS

INTRODUCTION 69

Chapter SIX: Scorpions in Captivity . . . 77

Chapter SEVEN: Breeding 81

Chapter EIGHT: Caring for Young Scorpions . 83

SPECIES ACCOUNTS

Deathstalker Scorpion *Androctonus amoreuxi* . 86
Blood Red Scorpion *Babycurus jacksoni* . . 87
Central American Bark Scorpion *Centruroides margaritatus* . 88
Striped Scorpion *Centruroides vittatus* . . . 89
Flat Rock Scorpion *Hadogenes troglodytes* . . 91
Desert Hairy Scorpion *Hadrurus spadix* . . 93
Tricolor Scorpion *Opistophthalmus wahlbergii* . 94
Emperor Scorpion *Pandinus imperator* . . 95
Israeli Gold Scorpion *Scorpio maurus palmatus* . 97
Stripe-tailed Scorpion *Vaejovis spinigerus* . . 99

SUGGESTED READING 101

TARANTULAS

Tarantulas and Scorpions

INTRODUCTION

The keeping of tarantulas as pets has progressed relatively quickly and relatively quietly in the United States. A hobby that effectively began in the late 1980s and exploded in the early 1990s has seemingly plateaued into a group of passionate and intensely focused professional and amateur tarantula keepers and breeders. Some keep a variety of tarantulas, but most seem to specialize in a certain group such as tree spiders, terrestrial

Adult female *Poecilotheria ornata*, one of the most striking members of the Genus *Poecilotheria*. Photo by Bill Korinek.

tarantulas, or baboon spiders. The number of specimens and species being kept is quite amazing.

As man spreads further into natural environments, not only do the vertebrates suffer, but the invertebrates are disappearing as well. Sadly, little protection is offered to tarantulas other than passive protection from land set aside for larger, high-profile species.

With this book, I hope to inspire new keepers. Included are the basics as well as helpful hints for successful husbandry and breeding that have helped keepers maintain some of the most exciting tarantulas and scorpions in captivity.

Russ Gurley

Chapter ONE: Tarantulas as Pets

Most members of the Genus Grammostola, such as this *G. aureostriata*, are quite docile and make wonderful pets. Photo by Russ Gurley.

Is an invertebrate pet right for you?

Other than hissing roaches, hermit crabs, millipedes, snails, stick and leaf insects, and some of the mantids, most invertebrate pets are best not handled. As with marine fish, the excitement comes in designing and creating an exotic jungle forest or an arid desert habitat in an enclosure in your home. Many of the tarantulas on the pet market and discussed in this book are very quick, stress easily, will jump and injure themselves, or may deliver a very painful bite. Some should only be kept by advanced keepers with plenty of experience with potentially dangerous animals. I suggest that a keeper starts small, with one or two of the hardiest and calmest species. This will allow him or her to gain the important experience required.

Read, question, explore, and weigh your realistic abilities, physically, financially, and emotionally, and if you decide to enter the world of tarantula keeping and breeding, this book should be helpful in beginning your journey.

Finding Tarantulas

Once a keeper has chosen a tarantula for a pet and has re-searched and designed an exciting enclosure that is both beautiful and correct (but also practical), it is time to begin your search for specimens.

Imported / Wild-caught Specimens

During the 1970s and 1980s, almost every captive tarantula in the pet trade was an imported, wild-caught animal. South American Pink-toed Tarantulas (*Avicularia avicularia*), Haitian Brown Tarantulas (*Phormictopes cancerides*), and the ultimate pet spider, the Red-kneed Tarantula (*Brachypelma smithi*) were the most commonly encountered. They were available at pet stores specializing in exotic species. In just twenty years, by the explosive 1990s, spiders and other invertebrates were being imported by the thousands. In addition to the influx of wild-caught specimens were shipments containing hundreds of captive-produced specimens from breeders in Europe.

Captive-hatched Specimens

Availability of captive-hatched tarantulas has increased dramati-cally as breeders continue to figure out the special needs and "triggers" to get their captive tarantulas to mate and successfully produce offspring. As the importation of wild-caught exotic pets continues to decline, the captive-hatched tarantulas will be needed to expand the hobby and to increase the numbers sought by ardent collectors. As these captive-hatched specimens are typically healthy, disease-free, and parasite-free, they are excel-lent animals to begin a collection.

Places to Find Tarantulas

Shows and Expos

There are reptile shows and expos all over the country. With excitement, we have noted an increase in the number of captive-produced tarantulas at these shows. Typically, the specimens offered at these shows are healthy, feeding well, and are excellent specimens to begin a hobby. At these shows you get the rare opportunity to hand-pick the tarantulas you want to purchase and you often have the opportunity to speak with an experienced keeper or breeder. When buying on-line or having tarantulas shipped to you, there is always the risk of receiving animals that are picked by someone who may not have your best interest at heart. Add to the savings of not having to pay shipping and the lack of stress placed on the animals from shipping and the shows and expos are an excellent opportunity to get some really nice tarantulas. The National Reptile Breeders Expo, the North American Reptile Breeders Conferences in Anaheim, Chicago, and Philadelphia, and the American Tarantula Society's annual conference are a keeper's best chance of finding large numbers of exciting invertebrates in the United States.

Tarantula Dealers

Some keepers are fortunate to have a local tarantula specialist near their home. Often, these keepers will welcome visitors to their facilities. In this situation, you get to see the breeder's setups and see his or her animals. You might learn some of their tricks, glean some experience and helpful hints from them, and often gain a new friend or colleague with whom to share ideas and offspring. You can find these keepers through a local invertebrate or herp society, a local university entomology department, ads in a reptile magazine, or on the Internet.

The Internet

The Internet has very quickly developed into a source of live animals. There are several extensive websites that offer classified ad sections where one can buy animals and plants as well as cages, supplies, food, and more. There have unfortunately been occasional problems with unscrupulous, faceless dealers. When buying this way, one doesn't get to see the animals or the facilities, and many of these Internet dealers are simply buying and reselling animals.   If you are careful and inquisitive, these Internet dealers can be a good source for tarantulas. When you contact dealers selling tarantulas, ask plenty of questions. These people want to sell you a live animal and keep you as a future customer so they should be willing to spend a little extra time with you. Make sure they are charging a fair price by looking around at what these tarantulas typically sell for in other ads and from other sources such as dealer price lists. Do your homework. Most will be willing to send you photos of the specific animal in which you are interested. Find out about their packing and shipping techniques. Make sure they sound legal, logical, and safe for the animal. If the seller is rude or unwilling to answer your questions, move on and count your blessings. Typically, these deals end up being the ones you regret.

Shipping and Transporting Tarantulas

There are concerns about shipping tarantulas, even with over-night delivery services. There are Styrofoam-lined boxes, disposable heat and cold packs, and most boxes can travel across the country in a day without a problem. We try to only ship and receive tarantulas from April to October and are careful during cold nights in winter and even more careful about shipping during hot days in summer.

Pet & Specialty Stores

As interest grows, more and more pet stores are offering tarantu-las for sale. Not only are they offering tarantulas, scorpions, and other invertebrates for pets, but they are exhibiting healthy animals in proper and inspiring setups. Many are offering correct advice and stocking the best equipment and supplies for their customers. Now, with the increased emphasis on the true needs of spiders and other invertebrates, shops are installing better invertebrate enclosures such as screen cages, cages with screened sides, vertically oriented terrariums, etc. They often offer to help construct exciting and inspired setups for customers. They are also using better food and prey items.

Though many continue to get a bad rap, these shops are literally the front line in our crusade to educate the general public about spiders, scorpions, mantids, and other invertebrates. As the first stop for most people searching for an "unusual" pet, pet shops have the unique ability to inspire a beginning keeper's first creative ideas and to offer proper procedures for setting up and caring for captive tarantulas.

Choosing A Specific Tarantula

When you discover a tarantula that you are interested in purchasing, begin by checking out the enclosure. If there is a water dish in the enclosure, check the water. It should be clean and free of dead insects. Make sure the cage is relatively free of feces. This will give you an idea of how often the tarantulas are cleaned and how much attention they are receiving. Most stores will allow their employees to toss in a cricket so you can see if the tarantula has a good feeding response.

Check the tarantula that has caught your eye. Make sure it has all legs (though most inverts will grow back missing or damaged legs over time as they molt), is free of injuries, bumps, etc. Bumps, lumps. or an assymetrical abdomen are usually signs of injury, disease, or a parasitic infection. Ask about any guarantee the seller might offer. Is this guarantee offered in writing? Remember, you are often stuck with your decision with no possibility of a refund. In fairness, the seller can't know the care you will offer and so can only guarantee the

For species that are sort of mid-range in aggression, such as this Zebra Tarantula, *Aphonopelma seemanni*, a firm grip with the legs held close to the body is recommended. Photo by Russ Gurley.

The King Baboon Spider, *Citharischius crawshayi*, is a large, beautiful spider, but it is very aggressive and will eagerly deliver a dangerous bite. Photo by Russ Gurley.

tarantula's current health.

As mentioned earlier, most tarantulas do best when kept similarly to tropical fish. A keeper should design and construct an interesting and creative enclosure, add the tarantula to the environment, and enjoy watching its behaviors with minimal interference and handling. Of course for those interested in breeding their tarantulas, some manipulation is required. Typically, males and females will be kept separately and one will need to be introduced into the other's environment after a period of conditioning. Most of the time, mating attempts will require the keeper to stay close by and protect the male, who is occasionally seen as not only a mate, but a meal.

Handling

For tarantulas that can be held safely, a handler should typically sit on the floor when handling his or her tarantula pet. If an animal jumps or falls (which is common), it will probably not receive any serious harm from this shorter distance. All children

This large vivarium with sliding glass front doors is a beautiful addition to any home. This enclosure features live plants and plenty of limbs and pieces of bark for the arboreal tarantula living there. Photo by Bill Korinek.

should be supervised and instructed on careful handling procedures. We suggest hermit crabs, hissing roaches, millipedes, or snails for children wanting an unusual but safe invertebrate pet. As with all pets, anyone who handles an animal should be sure to wash his or her hands after handling their pets.

Bites

Many of the tarantulas featured in this book are capable of delivering a painful and sometimes medically serious bite. Most American species of tarantulas will flick hairs from their abdomens as a defense mechanism. These hairs have microscopic barbs that irritate the mucous membranes of would be attackers. The baboon spiders of Africa rely on a venomous bite rather than flicking these hairs. Also, many of the arboreal tarantulas such as *Poecilotheria* species are fast and aggressive and will readily bite a careless keeper.

Chapter TWO: Housing

ADULTS

The Enclosure - Aquariums

Glass and acrylic aquariums can be excellent enclosures for tarantulas. We suggest a lower profile aquarium for terrestrial species and a vertically oriented enclosure for arboreal species. As they grow, they will of course need larger and larger enclosures. Typically, aquariums are relatively inexpensive, available in a variety of sizes, and visually interesting

Those keepers with large collections of spiders often use inexpensive plastic tubs and containers for their spiders. Photo by Russ Gurley.

when set up in a special part of a keeper's home. As they are clear, they work best for naturalistic enclosures as the plants growing within receive beneficial light more easily. Secure screen tops are available for these glass aquariums and are usually easy to find at local pet stores or on-line. Custom-made enclosures are also popular with keepers and can be built inexpensively from sheets of glass and plexiglass by a keeper that is proficient at this type of work.

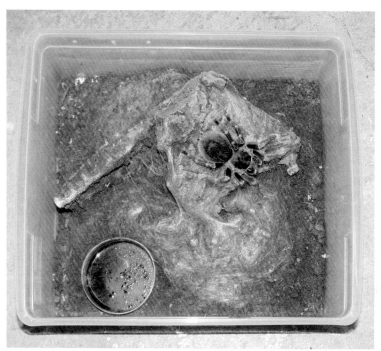

Substrate in a typical tarantula enclosure can be peat moss or a mixture of peat moss and sand. Photo by Russ Gurley.

Plastic Containers

Many tarantula keepers, especially those keeping a number of specimens, choose to keep their tarantulas in plastic shoe boxes, half-gallon and gallon jars, and large plastic tubs. These plastic containers are inexpensive and available in a wide variety of shapes and sizes at a local department store. They are typically semi-transparent and thus offer the spiders a more "secure" environment. Though they typically don't make attractive enclosures, they work well in larger collections as they are lightweight, easy to clean, hold in moisture well, and are inexpensive.

Naturalistic Enclosures

There is a growing movement in arachnoculture for the establishment of beautiful and elaborate naturalistic vivaria for invertebrates. The business of selling driftwood, moss, misting systems, colorful sand, and supplies is thriving. Many keepers are choosing naturalistic setups for their invertebrate pets. They feel, and I agree, that these naturalistic setups provide more interesting enclosures and may indeed hold the key to breeding some of the most frustrating species that have so far not cooperated in breeding consistently in captivity.

Substrate

Substrates are an important addition to the look and long-term health of a naturalistic setup. We choose to use a mixture of sand and peat moss for most of our invertebrates' enclosures. It is not only natural, it holds moisture well and the acidity of the peat seems to keep the development of mold or fungus in check.

There is some concern about tarantulas ingesting sand as they feed on crickets or small mice that wander their enclosure. We are not overly concerned and we feel that healthy spiders will simply pass these small amounts of sand with their feces. This ingestion of soil no doubt occurs in nature and seems to cause little or no problems. We do not use vermiculite as a substrate because it sticks to spiders and can swell if ingested, causing impaction and death.

Cage Decorations

Natural Shelters

There is no doubt that the addition of plants, driftwood, cork bark, stable rock piles, and other cage decorations is important in keeping captive tarantulas healthy and stimulated. These decorations can provide important hiding places, allowing the animal to feel secure and producing an area for safe molting and production of egg sacs. Cork bark tubes are ideal shelters for most terrestrial and arboreal tarantula species. Unfortunately, they are such excellent shelters that one rarely sees spiders when they have a tube in their enclosure. If you collect branches and other decorations from nature, be sure that they come from an area that is not sprayed with pesticides and that they are non-toxic.

Rocks and stones can be sterilized in a light bleach solution (one part bleach to five parts water) and then scrubbed with a soap-filled sponge and rinsed thoroughly before they are added to the enclosure.

Be sure that any cage decorations are firmly resting on the bottom of the enclosure and that the substrate is pushed up against the object. If a spider digs under a flat rock or other heavy object, the object might settle onto the spider and crush it against the bottom of the enclosure. This accident is fairly common in the collections of beginning hobbyists.

Man-made Shelters

There are a variety of reptile shelters on the market that can be easily converted to use with invertebrates. Some of the best are the natural-looking rock shelters and plastic or stoneware "caves". Some "thrifty" keepers also use upside-down butter dishes, whipped cream tubs, pieces of PVC pipe, clay pots, and other found objects. These shelters are typically easy to clean and sterilize (or inexpensive to discard) and are in sizes perfectly suited for invertebrate enclosures.

Live Plants

Live plants make excellent additions to the enclosures of arboreal species. They not only make attractive enhancements to the enclosure, they provide some moisture and humidity and provide natural resting places for your spiders. Hardy, shade-loving species work best. These include *Sanseveria* species, hardy bromeliads and "air plants", *Pothos* ivy, and English ivy, among others. Large terrestrial tarantulas such as *Theraphosa* and *Pamphobeteus* will typically travel around their enclosures and trample any plants in with them. Baboon spiders will tend to use plants as a basis for their webbing, but may also dig up the plants while excavating and modifying their environment. Unfortunately, many spiders will cover the plants so thickly with webbing that the plants die.

The artificial plants and other additions within this enclosure offer the tarantula places to hide, to climb, and to attach webbing to construct its shelter. Photo by Russ Gurley.

Artificial Plants

Plastic or silk artificial plants make a nice alternative to live plants in the spider enclosure. They add some color and places for shelter, breeding, and egg sac production. Also, webbing and trampling do not adversely affect their health.

Molting

As your tarantula grows, it will need to molt its old exoskeleton to grow. Spiderlings typically molt once a month and larger spiders will molt once or twice a year. Many species of tarantulas will darken in color in the weeks prior to molting. They will also usually stop feeding and become somewhat lethargic. At this time, a keeper needs to make sure that the area underneath the spider's shelter is humid. This humidity will help ensure that a spider molts successfully. Many captive spiders become injured or die during molting due to inadequate conditions in their enclosures.

Molting is a very delicate process. In the days before molting, a layer of fluid and air will form between the spider's old skin and the new skin developing underneath. When the molt is imminent, a spider

This freshly molted *Megaphobema mesomelas* will be somewhat delicate and susceptible to injury for the day or two following its molt. Photo by Bill Korinek.

will usually web the entrance to its burrow or shelter closed. This keeps unwanted insects out and helps seal some of the humidity inside. The spider will lay down a carpet of white silk web and will typically position itself in the center of this web. The spider will flip over on its back and the process of molting will begin.

The spider will begin gently pumping its legs, pumping blood out into its body. The outer old skin will begin to split, first along the carapace and then along the midline of the abdomen. The spider will begin slowly pulling itself out of the split in the carapace. Once it has removed its cephalothorax and abdomen, it will then gently pull each leg out of the old skin. This is a delicate process and if humidity is too low, a leg may get stuck inside the old skin and the spider may lose that leg. In some cases, a spider will lose a leg or two. At other times, the spider may break off a leg and bleed to death or become stuck in the dry skin and die.

After molting, the spider will be soft and extremely vulnerable to any poking or prodding. Crickets will attack and kill spiders during this time. It will take a while for the spider's new skin and even fangs to harden up. Spiderlings may harden up and begin moving and even eating within a day, while adults may take up to a week to get back to normal.

Molting allows a spider to replace lost legs, to replace the urticating hairs that the spider has kicked off its abdomen, and the females become virgins again.

Chapter THREE: Feeding

Feeding tarantulas is simple and straightforward. They are going to need a diet consisting of a variety of live prey: Crickets, roaches, grasshoppers, mealworms, waxworms, flies, moths, and occasionally pink or fuzzy mice may be required.

Small tarantulas will need to be fed small insects and as they grow, they will require larger prey items. We

Tarantulas of all sizes will eagerly feed on crickets. This is a spiderling *Avicularia versicolor*. Photo by Bill Korinek.

feel strongly that all captive invertebrates should be offered a wide variety of prey rather than just feeding those that are easiest to find at the local pet store or bait shop. Many keepers, especially those with large collections, also raise several types of feeder prey in connection with their invertebrate collections. This insures that they have a constant supply of prey items that are well-fed and healthy and provide the best nutrients for their captive pets. Some feeders such as crickets, fruit flies, and wax moths are especially time-consuming to produce.

Feeding your tarantulas prey items that have been gut-loaded, or fed a very nutritious diet, in the hours or days before feeding them to your pets is the ideal situation. Gut-loading of crickets, roaches, and mealworms with healthy food is an important part of

A simple setup in a plastic tub with mulch, egg cartons, a water container, and an ample supply of fruits, vegetables, and grains, is the ideal way to culture crickets and roaches for your pet tarantulas. Photo by Russ Gurley.

feeding your tarantulas and scorpions. By feeding (gut-loading) your prey items a healthy diet, these nutrients are transferred to your pet inverts.

Gut-loading Crickets and Roaches

Crickets and roaches should be filled with a healthy meal in the hours before you feed them to your invertebrates. We recommend adding a small pile of 5-6 of the items below to the cricket or roach enclosure once a week:

Romaine lettuce
Greenleaf and redleaf lettuce
Mustard greens
Collard greens
Dandelions
Green beans
Yellow squash
Zucchini

Tarantulas and Scorpions

Carrots (shredded)
Sweet potato (shredded)
Apple slices
Orange slices
Pears
Cantaloupe
Mango
Papaya
Oats
Wheat germ
Corn flour
Rice cereal baby food
Powdered milk
Sunflower seeds (unsalted)
Bee pollen
Tropical fish flakes
Spirulina flakes

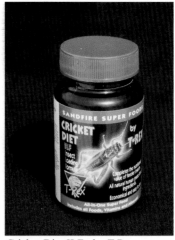

Cricket Diet ILF® by T-Rex products is an ideal addition to the gut-loading diet for feeder insects.

Water

Depending on its life in nature, a tarantula will require anything from a damp, humid enclosure with frequent mistings to an arid habitat with low ambient humidity and only its prey item as a source of water. For more tropical enclosures, daily or twice daily spraying may be required. For forest enclosures, less frequent spraying or a drip system may be enough.

Chapter FOUR: Breeding

Mating can be a very tedious affair for male tarantulas. If females are unreceptive, the male can quite easily become a meal rather than a suitor. Photo by Bill Korinek.

Sexual Dimorphism

Though they look alike for most of their lives, both sexes of tarantulas exist. Typically males and females look very similar until the male's final molt. At this molt, males emerge looking quite differently from females. Some have long, furry legs, darker coloration, or even a metallic tint. They will have swollen palpal bulbs and many will have tibial spurs.

To begin a captive breeding program, typically, six or seven young spiders are purchased to form the basis of a breeding group. Of course the breeders that are producing large numbers of spiderlings begin with much larger numbers to ensure that they end up with plenty of mature females in their collection. As already mentioned, whenever possible, a keeper should purchase as many unrelated spiders as possible. This will help ensure that genetic diversity of their breeding program is high. This can usually be achieved by purchasing small groups of spiderlings from a variety of sources. However, one must keep in mind that many tarantula dealers frequently buy and sell each other off-spring.

SEXING TARANTULAS

Visual Inspection

When a male reaches his final molt, he will change quite dramatically physically. His coloration may intensify and his legs may be elongated and sculpted or unusually furry. He will also be sporting new or modified parts in the form of swollen palpal bulbs, enlarged pedipalps, and tibial hooks (in most species) for managing the fangs of would-be mates. As there are obvious benefits to knowing beforehand if you have females or males (and the ratio of each), methods for sexing tarantulas before this final molt have become very useful tools for the tarantula breeder. I would like to present a couple of methods in use by tarantula breeders.

External Characteristics

Some species of tarantulas can be sexed visually by inspecting their anatomy. This method is not fool-proof and is not reliable on some species and some smaller specimens.

The epigynal plate, anterior to the epigastric furrow, tends to be more pronounced and raised in female spiders. In males, this plate is smaller and flatter. The shape of the plate is variable, but in general, the plate of females is more triangular.

Observing the Exuvium

Once a spider reaches roughly 2-3" (depending on the species) it can often be sexed by making a visual inspection of its molted skin, or exuvium. The reason this can be done is that a tarantula sheds the lining of its sexual organs when it molts. In this method, a keeper must carefully spread the recently molted skin on a flat surface. (A dried skin can be used, but should be

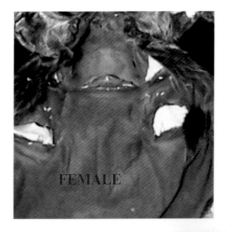

The flowery spermathecae can be seen emerging from the anterior portion of the epigastric furrow in the female's shed exuvium (just anterior to the white book lungs).

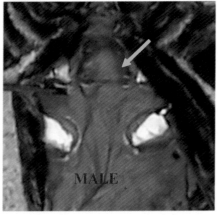

softened with water or alcohol beforehand.) With a 3x to 10x magnifier, look between the anterior pair of book lungs (white structures). There you will see a line of folded skin, the epigastric furrow. Depending on the species, a female tarantula will have leafy structures in this area. These structures are the spermathecae. The female stores the males' sperm in this area after mating and before egg-laying. Males do not have the spermathecae, so their epigastric furrow will simply be a line of tissue.

Once Sex is Determined

Typically, once sex is determined, a keeper's goal is to raise the females to an adult size as quickly and healthily as possible.

They should be kept warm and fed a variety of prey items once or twice a week. Some species are slow-growing and maturity can take anywhere from a couple of years to as many as six or seven years, depending on the species.

When raised from a group of spiderlings, males typically mature much earlier than females. This is no doubt an evolutionary mode that keeps inbreeding to a minimum. (If males mature long before similarly aged females mature, they will be forced to search for females from another "source" than the locally available females which might be their siblings.)

Some keepers have experimented with pushing one group of spiderlings to get them to mature early and keeping one group fed on a maintenance schedule. Their hope has been to get a female to mature early and to have a male mature late, giving them a pair that is breedable from the same group of spiderlings. Most breeders today know the value of producing spiderlings from a breeding of two healthy, unrelated adult spiders. Unfortunately, for many of the rarest species there are so few specimens around.

A series of tapping and vibrating precedes mating in the Giant Colombian Tarantula, *Megaphobema robustum*. Photo by Bill Korinek.

Luckily, there is good communication between most spider breeders and often keepers advertise which females they have that are ready to breed and they often let other keepers know when they have a newly molted male that is ready for breeding.

Many of these connections lead to breeding loans. In these loans, two keepers reach an agreement about the future breeding of their two spiders. This agreement is usually a 50/50 split of any spiderlings produced from a successful mating. In these loans, typically the male travels to the female. The males, though delicate, are seen as "expendable" compared to the female who is more likely established in an enclosure with a safe, secure burrow or shelter and is heavy, having been well-conditioned for potential breeding. There are always dangers when shipping animals and losing a female because of a dropped box or mishandled shipment is not good.

When a male is shipped to a keeper, there is a chance that the male is old and may be unable to breed if he has not recently reached his final molt or if he has not built a sperm web. A keeper can usually spot an old male by a bald spot on its abdomen, a shriveled abdomen, or it may be dully colored or lethargic. These are all signs that the male has been around for a while and may be too old to breed successfully.

Pairing

Breeding tarantulas seems like it would be fairly simple. Indeed the basics are well-publicized. One simply takes a "charged" male tarantula and introduces him into the female's enclosure. However, the complexities of breeding tarantulas are myriad. First, a keeper must wait for a young spider to grow into adulthood. On his penultimate, or final, molt, a mature male will become equipped with some form of breeding apparatus. For many species, this will include swollen pedipalps, hooks on their front legs, and other anatomical changes. After a period of time that varies by species, the male with produce a sperm web, usually in the form of a small web platform. He will deposit sperm onto this web and then will dip his palpal bulbs into the fluid ("charging up"). He will use this fluid to fertilize a willing

The actual act of mating, as in this pair of *Ephebopus cyanognathus*, is typically the culmination of complicated courtship behaviors. Photo by Russ Gurley.

(or relatively willing) female partner.

Once introduced into the female's enclosure, the male will typically vibrate or tap his legs to signal to the female that he is a willing mate and not a prey item. This is obviously the most dangerous time for a male spider and many would-be breedings are lost in a flash as a female attacks and kills the male. This can often be avoided if the female is kept well-fed and if the female is post-shed * and in what the keeper would consider to be the most "receptive state" as possible.

* If a female sheds soon after mating, she will slough the interior lining of her spermathecae and thus lose any sperm being stored there. Thus, it is important that a female is bred soon after a shed or at a time when one is not imminent. A breeder who keeps good records of shedding dates will have a much better chance of successful pairings.

The female will typically approach the male when they are introduced. Those males with hooks on their first pair of legs will raise these legs up quickly and hook the female's fangs to hold them back and to raise the female's body, giving him access to the female's genital opening, or epigyne. He inserts the palpal bulb into this area at the underside of the female's abdomen and discharges the fluid to fertilize the female. The male will typically drum on the female's abdomen or underside of her cephalothorax to placate her further during the process. After mating, which may take from a minute or two to as long as thirty minutes, the male will release the female's fangs with his hooks and he will either carefully back away or race off.

In a small enclosure, the female may grab a male as he reaches the end of the enclosure. Some more docile species will allow the male to live and may even mate more times in the following days.

A spider breeder will usually leave the top off of the enclosure during a pairing attempt. This allows the breeder to keep close watch on the happenings and also allows the male an escape route once mating has occurred.

One of the most frustrating times for a tarantula breeder are when he or she sees several "by the book" pairings and an egg sac never arrives. There are many theories about why females fail to produce egg sacs. The most likely of these include females that are well-fed, but not fully conditioned. We are experimenting with a three-month cooling/resting stage for females prior to pairing and also after pairing to see if this affects the production of viable egg sacs.

Some other tips for successful breeding of many species of tarantulas include:

The Enclosure

For successful breeding, females of many terrestrial species seem to require a deep burrow. This can be achieved by simply starting a burrow for the spider in deep substrate and allowing it to dig further and reinforce the burrow with its own webbing. Another method is to create an artificial burrow via a plastic tube such as those used for gerbils or hamsters. Burrows can also be produced from a piece of clay pipe or one can be carved in a large piece of furniture foam which can be inserted into the enclosure. There are a wide variety of " burrows" in use by keepers

A deep, secure burrow seems to be an important part in the successful captive reproduction of many terrestrial species of tarantulas such as this Chinese Golden Earth Tiger, *Haplopelma schmidti*. Photo by Bill Korinek.

worldwide and there are no limits to what can be created. The burrow not only provides the spider with a more natural home and less stressful microhabitat, it also provides the female spiders a safe, stress-free place to produce and to nurture an egg sac. Those breeders who have been most successful in breeding terrestrial species, commonly provide their adult female spiders burrows in which to live.

Conditioning

A keeper who keeps his or her spiders healthy (not obese) is more likely to produce viable eggsacs with higher hatching rates. As with most captive animals, overfeeding and obesity are common. It appears that the same can be said for captive spiders too. Often keepers enjoy seeing their captives eat and this leads to overfeeding. We feel that by feeding an adult spider once a week and feeding a variety of healthy prey items, a spider can be kept healthier and is more likely to live longer and produce more and healthier spiderlings when bred.

"Seasonal" Changes

As our knowledge of the keeping and breeding of tarantulas in captivity improves, it appears that manipulation of environmental temperature can be important to stimulate breeding in spiders from temperate areas. Some of the more tropical tree spiders may also benefit greatly from a mild cooling period and by a change in seasons from dry to wet. We feel that a keeper should investigate the natural environment for each species he or she keeps. A weather report for each month of the year can be found on-line. This weather information can give the keeper an idea of the ideal situation for a captive specimen and what triggers might stimulate members of this species to reproduce in captivity.

THE EGG SAC

After a successful pairing, the female tarantula now has sperm stored in her spermathecae, or sperm storing organs. These spermathecae are pockets located just under the female's genital opening. After a varying period of time, the female will be ready to lay her eggs. The process of egg-laying is similar for most species. The female will first lay out a small carpet of thick, white web. Upon this web carpet she will lay from 25 to as many as 2,000 small eggs. The eggs are fertilized by the male's sperm as they pass the spermathecae. Once she has completed the egg-laying, she will pull the corners of the web carpet into a small packet containing the eggs.

This egg sac prevents predators from getting to the eggs and also helps seal in humidity. Depending on the species, the egg sac is either suspended within the burrow or a tube web or is trans-formed into a loose ball-shaped package that is carried around by the female. In these cases, the egg sac is grasped firmly in her chelicerae, or fangs, and is manipulated by the first pair of legs.

This female Goliath Bird-eating Spider, *Theraphosa blondi,* will aggressively protect her newly produced egg sac. Photo by Bill Korinek.

Those species that carry their egg sac around will often move the egg sac in and out of the burrow or tube web to expose the eggs to the proper warmth and humidity within the enclosure.

Most species will only lay one egg sac after a successful pairing. Occasionally, a female will lay two clutches of fertile eggs. When a female tarantula molts, she sheds the lining of the spermathecae, losing the sperm stored there, and thus loses the ability to produce fertile eggs.

To Take or Not to Take the Egg Sac

Most successful spider breeders will allow a female to carry her egg sac for a month to six weeks and then will take the egg sac away from her.

Occasionally, when an enclosure is in a room with lots of traffic or if the enclosure is bumped, a female may become stressed and eat the egg sac and eggs within. This is a frustrating experience and after several losses, a breeder often decides to take the egg sac and incubate it away from the female.

The key to incubating an egg sac is having a system in place that has worked successfully. We have used a simple setup using two large plastic tubs. This incubation setup has worked for a wide range of species and is simple and inexpensive. First, a keeper should fill a 6" to 8" diameter deli tub with 1-2" of damp peat moss (mix water and peat moss together and squeeze the excess water out). Over this tub is placed another tub from which the bottom has been removed with a sharp knife. On this top tub is placed a sheet of cheese cloth which is secured tightly with a rubber band or wire. The egg sac sits gently on the piece of cheese cloth, suspended above the humid substrate. This allows the egg sac to be kept humid, but keeps it up off the substrate to prevent it from getting too damp and getting moldy.

This setup is placed in a warm room (75° to 80° F) in a larger plastic sweater box with plenty of ventilation holes for the remaining incubation time. In the first weeks, the egg sac is gently rolled by the keeper six to eight times per day to keep the eggs within from clumping together.

The egg sac should not be taken from a female while the eggs are very recently laid. When the eggs are laid, the female discharges a viscous liquid that

The "two tub" egg sac incubation system is being used by the top tarantula breeders in the United States. Photo by Russ Gurley.

Tarantulas and Scorpions

coats the eggs, providing them with humidity and probably containing substances that dissuade any would be predatory flies and other small attackers. The female tarantula gently rolls the egg sac and the eggs within while she carries it with her. This rolling prevents the liquid and egg mass from sticking together and clumping. If an egg sac is taken too soon and the egg sac is placed on

Small, clean vials with tight-fitting ventilated lids make the ideal enclosures for small spiderlings. Photo by Russ Gurley.

the cheese cloth, the eggs and liquid will clump together, killing the developing eggs. Once the small nymphs begin to develop, this liquid is absorbed and will no longer cause the small spiders to clump together. An experienced breeder will be able to tell from the changing shape of the egg sac and by the feel of the small nymphs inside. The egg sac will feel looser and the ball of nymphs will be more spread out within the egg sac, telling the keeper that the nymphs are moving around somewhat. It is at this time that the keeper can remove the egg sac to a support structure to carefully watch the further development of the nymphs and to prevent a female, especially of a very rare species, from destroying what has been months or even years of hard work and preparation.

The nymphs of tarantulas look like light-colored peas with small, stubby legs. After a period of time, they will molt and look more like small, mobile spiders. Nymphs of *Avicularia* species and several other Genera will remain within the egg sac until they molt their skins for the first time. At this time they will emerge from the egg sac and spend some time together. Once they are mobile and ready to begin feeding, they will usually disperse into the enclosure and seek out a small space to begin their daily routine of waiting for food.

Spiderlings of many species will begin cannibalizing their siblings. They should soon be removed to small, individual containers.

Chapter FIVE: Caring for Spiderlings

This small Brazilian Purple Tarantula, *Iridopelma hirsutum*, can be gently moved by letting it climb from the hand to its enclosure. Photo by Russ Gurley.

Enclosures

There are a wide range of enclosures that are suitable for raising spiderlings. Most keepers use small plastic vials which can be ordered through a variety of suppliers. These small enclosures let the spiderling find and capture prey more easily. The plastic lids will need to fit tightly and will need to be perforated with small air holes to provide oxygen for the spiderling that will live inside. Some keepers use small condiment containers, deli cups, and other small plastic containers. Of course the type of spiderling being kept will dictate the enclosure that a keeper needs to establish. Arboreal species do well in taller vials and terrestrial species do better in shorter containers that offer more "floor space".

Plastic Containers

Plastic containers such as vials, deli cups, sandwich tubs, shoe boxes, and pint jars are available to provide a variety of sizes and shapes for spiderling enclosures. They are typically inexpensive and easy to find at any local department store. When used for keeping invertebrates, it is best to modify the lids or sides of these containers to allow airflow. For spiderlings, these air holes need to be quite small and can be made with a small, sharp nail, a needle, or other tool.

Substrate

Young spiders are easier kept on a "light-weight" substrate. We keep most burrowing species in vials or small containers filled about halfway with peat moss with a pinch of sphagnum moss on top. This moss "plug" will give the spiderling a place to attach some webs and will help hold in moisture. The moss is easy to moisten and dries out more naturally. The peat and sphagnum will also lighten in color when dry, giving the keeper an idea of when water needs to be added to the vial. Heavier substrates such as sand can settle onto the spiderling if the vial is bumped or jostled, killing the spiderling. Many keepers use paper towel as a substrate for spiderlings. It is both easy to clean and is slow to support the growth of mold and fungus which kills many captive spiderlings. It does, however, dry out quickly and a keeper must be diligent to not let the inside of the container become too dry. Dehydration kills many spiderlings.

Feeding

For their first few meals, spiderlings will feed on small crickets and wingless fruit flies once or twice a week. Most keepers add 3-4 small crickets per vial and spray a small spritz of water in the vial at the same time. This ensures that the substrate is moistened and establishes a schedule for feeding and watering the small spider. If they are kept warm and fed well, spiderlings will molt once a month. A 3" x 5" note card is an easy way to make notes such as feeding dates, molting dates, and other information.

Be sure not to feed your spiderling when it is freshly molted or will molt soon as crickets are quite ravenous and will eat a spiderling if it is in a weakened state.

Again, small enclosures allow easier access to prey items and tarantulas are used to living in burrows and tube webs, so cramped spaces are their natural living space.

If a large number of spiderlings are hatched, many keepers will keep 5 to 15 spiderlings in a single enclosure. They can usually be kept together until cannibalism becomes an issue and they will then need to be separated into individual containers.

Moisture

Spiderlings will die fairly quickly if they are allowed to overheat or dry out. A keeper should lightly mist the inside of the small containers once a week so that the substrate becomes moist, but not waterlogged. (Growth of mold and fungus within the enclosed space will also kill a small spider.) We have found that a light spray once a week works well and allows the enclosure's substrate to dry out inbetween mistings. The spiderlings typically get any additional moisture they need from their prey items.

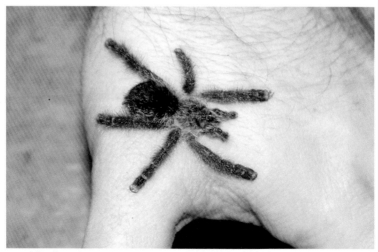

Humidity is an important part of a pet tarantula's life. This *Avicularia* lost two of its legs during molting due to an enclosure that was too dry. Photo by Russ Gurley.

Molting

One of the most dangerous times for a tarantula in captivity is during molting. A tarantula will molt more often when small and less often as it reaches adulthood. During each of these molts, a keeper must be sure that a proper molting area is available and that the humdity level of this area is within acceptable limits. If the area is too dry, the molting skin may stick to the spider as it pulls itself free. In these cases, the tarantula may lose a leg or two, may lose a leg and bleed to death, or may become stuck in its molt and die. One of the most common causes of death in captive spiders is death from getting stuck in a molt or injuries related to an improper molt.

Injuries

Another common cause of death in tarantulas is from injuries received from a fall. Bulky adult specimens of terrestrial species are most likely to receive life-threatening injuries when dropped. Small specimens and some of the more slender arboreal species are more likely to survive a fall.

When a large spider falls, its skin may rupture causing an injury that leads to extensive bleeding. There are reports of keepers repairing these injuries using Band-aid® Liquid Bandage or surgical glue, but unless the damage is minor (such as damage to a leg), the spider will usually die.

THE FUTURE

Enjoy your spiders! It is miraculous that some of them, so incredibly small as spiderlings, grow into the large, aggressive, hairy beasts that we enjoy so much.

Habitats across the planet are under attack. With the destruction of forests comes the loss of spiders. The spiders in captive breeding programs across the planet are potentially the only source for future populations of tarantulas, both for the hobby and for possible reintroduction into "recovered" ecosystems.

Keeping healthy spiders in creative enclosures will not only inspire others who visit your collection, but will also provide the potential for breeding and production of healthy spiderlings to share with other keepers.

I suggest that you make contact with other keepers - trade, sell, buy, and enjoy the unusual hobby of spider keeping and breeding!

SPECIES ACCOUNTS

Brazilian White-banded Tarantula
Acanthoscurria geniculata
Brazil

Comments: This large terrestrial species (6" to 7.5") from the forests of northern Brazil is striking. Its heavy body is covered with dark hair and it has thick black legs with beautiful white patches on each. This species is quite nervous and aggressive and should not be handled.

Photo by Bill Korinek.

Enclosure: As a large forest species, *A. geniculata* should be offered a large enclosure with plenty of extra floor space. The enclosure should be kept warm (78° to 82° F) and the substrate should consist of a thick (4") layer of lightly dampened sand and peat moss with a covering of sphagnum moss, cypress mulch, or leaves. A complex enclosure is not necessary, but this species will need a section of curved cork bark or other shelter, especially during molting. Misting with water once a week is recommended and a small water dish should be kept in the enclosure at all times.

Feeding: This species should be offered a variety of larger prey once or twice each week. They will feed aggressively on adult crickets, grasshoppers, roaches, and will even eat lizards and small mice. (Many keepers do not feed vertebrate prey to their tarantulas. If small mice or lizards are offered, for instance to condition females for breeding, we suggest only once or twice over the period of a month or two.)

Sexing: Male *A. geniculata* are slender and long-legged following their final molt. They also lose the striking white leg patches and have only faint bands of pale color at the leg joints.

Breeding: A mature male, once he has successfully completed a sperm web, should be carefully introduced into the female's enclosure. The male can be protected with a piece of cardboard or other tool if he is to be used for further breeding attempts. One should be careful not to distress or "anger" a female while manipulating the male or to cause the male to race over to the female while escaping your prodding. Calm and quiet is called for. Once mating occurs, the keeper should continue feeding the female a variety of prey items in hopes of receiving a large egg sac full of fertile eggs.

Additional Notes: Spiderling *A. geniculata* thrive in small enclosures that offer 1" to 2" of lightly dampened peat moss substrate and a piece of sphagnum moss for them to attach webs and build a secure burrow for attacking prey. They grow quickly and make beautiful display animals.

Bloodleg Tarantula
Aphonopelma bicoloratum
Mexico

Comments: *A. bicoloratum* is a beautiful spider that is hardy and long-lived when set up properly. It has been bred in captivity and is occasionally available from terrestrial tarantula enthusiasts as captive-produced spiderlings.

Enclosure: Like many of the tarantulas of southern Mexico, *A. bicoloratum* can be kept warm and dry, but they will need a somewhat cooler and more humid burrow or shelter for finding refuge from the heat and for a proper molting area. We suggest an enclosure temperature ranging from 78° to 82° F. The substrate can be dry sand and peat moss. A keeper should mist the enclosure lightly once a week and

Photo by Bill Korinek.

a shallow water dish should be available at all times. A shelter and a live *Aloe* or *Sanseveria* (or artificial plant) can be added

Feeding: This medium-sized spider will be an aggressive feeder on a variety of insect prey including adult crickets, grasshoppers, roaches, and waxworms.

Breeding: The Bloodleg Tarantula can be bred in captivity if certain pre-mating conditions are maintained. The females seem to benefit from a cooling period of a couple of months prior to mating. Once a mature male is produced, and he makes a sperm web, he should be introduced into the female's enclosure. He will approach the female's shelter cautiously, tapping and vibrating his legs. The female will be "lured" out of her burrow or shelter and the male will typically lunge forward to use his hooks to hold the female's chelicerae and to push her into an almost upright position to give himself access to the female's epigyne for mating. The male will insert either the left palpal bulb, right palpal bulb, or both alternately into the female's epigyne and inject the fertilizing fluid into this area. If this pairing is successful, the female will produce an egg sac in the following weeks. This species produces medium-sized egg sacs, usually containing 200 babies. A mature male can be introduced to multiple females or can be reintroduced to a female to enhance the possibilities of a successful pairing. Typically, the male will die in the weeks following a successful mating.

Additional Notes: This, and most slow-growing species, will rarely if ever repopulate the areas from which they are removed. It is certain that the specimens in captivity are very important for increasing numbers of this species to provide specimens for future hobbyists and perhaps for future reintroduction back into its former habitat.

Pink-toed Tarantula
Avicularia avicularia
Brazil, Trinidad, Martinique, Guyana, French Guiana, Suriname, Venezuela, and throughout the Amazon Basin

Comments: The Pink-toed Tarantula is one of the most re-warding species to keep in captivity. It is docile and easily handled and is hardy and entertaining if kept properly.

Enclosure: The Pink-toed Tarantula should be kept in a large, vertically oriented enclosure. A modified aquarium or tall plastic storage tub will work well. As with the other species of *Avicularia*, *A. avicularia* requires the unique combination of high humidity and plenty of ventilation. This combination can be somewhat difficult to provide in captivity. We suggest keeping the enclosure dry and spraying lightly with water every few days. The cage should be allowed to dry out in-between

Photo by Bill Korinek.

misting. We recommend that a keeper maintains 78° to 82° F. By keeping several live plants within the enclosure, a keeper can add to the humidity and keep these spiders in the suggested 65 to 75% humidity range. These plants can be placed within the enclosure still in their pots or can be planted in the deep substrate. These live plants will not only provide some more humidity, they will provide excellent areas for breeding and egg-laying. We suggest adding one or two shallow water dishes in the enclosure, depending on the outside conditions of the room in which the enclosure is located.

Feeding: This medium-sized spider will be an aggressive feeder on a variety of insect prey including adult crickets, grasshoppers, roaches, and especially flying insects such as wax moths. In nature, they feed on small lizards such as *Anolis* species, but we do not typically feed vertebrate prey to our invertebrates.

Sexing: Male *A. avicularia* have long, furry legs and the male is an overall black spider with more vividly colored "toes" than adult females.

Breeding: Adult males should be carefully introduced into the female's enclosure after he has produced a sperm web. The male can be protected with a piece of cardboard or other tool if he is to be used for further breeding attempts. Once mating occurs, the female should be fed a variety of prey on a more frequent schedule.

Martinique Pink-toed Tree Spider
Avicularia versicolor
Martinique

Comments: The Martinique Pink-toed Tree Spider is one of the most beautiful species kept in captivity. It is not as docile and easily handled as other species of *Avicularia*. This species requires the unique combination of high humidity and lots of ventilation. This combination can be somewhat difficult to provide in captivity.

Enclosure:
The Martinique Pink-toed Tree Spider should be kept in a large, vertically oriented enclosure. Modified aquaria or large plastic tubs work best. If the enclosure becomes too dry, the spiders will not do well. We recommend that a keeper maintains 78° to

Photo by Bill Korinek.

82° F with a humidity level of 75 to 85% for this species. One way to overcome the dilemma of high humidity and high ventilation is to have plenty of venitlation and to use a deep (4 to 5") substrate of damp sand and peat moss and provide several live plants within the enclosure. These plants can be placed within the enclosure still in their pots or can be planted in the deep substrate. These live plants will not only provide excellent places for the spiders to establish homes, they will provide areas for breeding and egg-laying. We suggest adding one or two shallow water dishes and misting the entire enclosure once a day to every other day, depending on the outside conditions of the room in which the enclosure is located. The cage should be allowed to dry out inbetween mistings.

Feeding: This species feeds aggressively on crickets and are especially fond of small, flying insects such as flies and moths. Spiderlings and juveniles will require very small crickets and even wingless fruit flies as prey.

Sexing: Males are darker, more slender, and long-legged.

Breeding: Adult males should be carefully introduced into the female's enclosure after he has produced a sperm web. Once mating occurs, the female should be fed a variety of prey items.

Curly-haired Tarantula
Brachypelma albopilosum
Costa Rica

Comments: The Curly-haired Tarantulas make wonderful pets as they are docile, hardy, and very long-lived.

Enclosure: This species lives in burrows in nature, but will do well in an enclosure that offers them 3-4" of sand and peat moss substrate and a sturdy shelter in the form of a curved piece of cork bark.

Photo by Bill Korinek.

They prefer temperatures in the 75° to 85° F range and a humidity in the 60 to 70% range. They can be kept dry if they are sprayed weekly and a shallow dish of water is kept in their enclosure at all times.

Feeding: Curly-hairs feed on adult crickets, grasshoppers, and roaches.

Sexing: Adult male *B. albopilosum* go through a rather dramatic sexually dimorphic change on their final, or penultimate, molt. They change from a stocky brown spider to a slender spider with long spindly legs and an overall black coloration. They also have tibial hooks and swollen palpal bulbs, used for breeding.

Breeding: The Curly-haired Tarantula can be bred in captivity if certain pre-mating conditions are maintained. The females seem to benefit from a cooling period of a couple of months prior to mating. Once a mature male is produced, and he makes a sperm web, he should be introduced into the female's enclosure. He will approach the female's shelter cautiously, tapping and vibrating his legs. The female will be "lured" out of her burrow or shelter and the male will typically lunge forward to use his hooks to hold the female's chelicerae and to push her into an almost upright position for mating. If fertilized, the female will produce an egg sac in the following weeks. This species produces large egg sacs, usually containing in excess of 400 babies.

Additional Notes: When disturbed or startled, this species will flick a substantial number of irritating urticating hairs from its abdomen. Highly stressed animals will have "bald patches" on their abdomens as remnants of this behavior.

Mexican Red-legged Tarantula
Brachypelma emilia
Mexico and Panama

Comments: *B. emilia* reaches 4 to 5" in total length. It is one of the more colorful members of the Genus. It has a pinkish carapace and a black abdomen covered with red hairs.

Enclo-sure: The ideal *B. emilia* enclosure offers spiders a

Photo by Russ Gurley.

deep substrate in which they can burrow. A sturdy shelter will give them a place to dig this burrow so that they may have a cooler, more humid area for shelter and molting. In a ten-gallon terrarium, we use a substrate consisting of 6-8" of damp sand and peat mixture. We add a cork bark or slate

cave-like shelter and try to add at least one sturdy live plant to the enclosure. We try to maintain the temperature at 78° to 82° F and a humidity of 60 to 70%. If kept drier, a keeper should spray the enclosure once or twice a week. A shallow water dish should be available at all times.

Feeding: *B. emilia* feed well on crickets, grasshoppers, roaches, and mealworms and large females being conditioned for mating and egg production will take an occasional (once a month) pink mouse.

Sexing: Male *B. emilia* are darker overall and have a metallic sheen to their carapace. They are also thinner in appearance.

Breeding: Breeding is typical for *Brachypelma* species. Tapping and vibrating is exchanged and the female will usually rush up to the male. If she is receptive, mating takes place without incident. If she is unreceptive, she will usually not signal by tapping and may quickly kill the male unless a keeper is diligent. Successful matings lead to egg sacs in 10 to 12 weeks. An egg sac typically contains 300 to 600 eggs and incubates for five to six weeks.

Additional Notes: Once abundant in the hobby, *B. emilia* has become quite rare due to a lack of successful captive breedings.

Mexican Red-kneed Tarantula
Brachypelma smithi
Mexico

Comments: *B. smithi* remains the most popular pet spider in the hobby. This medium-sized terrestrial species is protected by CITES (Convention on International Trade of Endangered Species). *B. smithi* is a docile species and usually makes an

Photo by Russ Gurley.

excellent animal for novice keepers.

Enclosure: As *B. smithi* lives in deep burrows along the Pacific Coast of Mexico, a captive enclosure should try to mimic these conditions. In a ten-gallon enclosure, we use a deep (6 to 8") substrate composed of a mixture of slightly damp sand and peat moss (50/50 ratio). Add a shelter at one end and begin a burrow under this shelter. The spider will typically continue the excavation. We keep the enclosure in the 78° to 82° F range. A heat lamp over an area just outside the entrance to the burrow will provide the needed warmth. (In nature, this species will often "bask" at the entrance of its burrows and will even drag egg sacs to this area to warm the developing embryos within the egg sac.)

Feeding: The Red-kneed Tarantula will feed aggressively on large insects and an occasional small mouse.

Sexing: Males of *B. smithi* are long-legged and dark overall with more vivid red-orange on the legs.

Breeding: *B. smithi* can be bred in captivity if certain pre-mating conditions are maintained. The females seem to benefit from a cooling period of a couple of months prior to mating.

Once a mature male is produced, and he makes a sperm web, he should be introduced into the female's enclosure. He will approach the female's shelter cautiously, tapping and vibrating his legs. The female will be "lured" out of her burrow or shelter and the male will typically lunge forward to use his hooks to hold the female's chelicerae and to push her into an almost upright position for mating. This species produces large egg sacs, usually containing in excess of 500 babies. A mature male can be introduced to multiple females or can be reintroduced to a female to enhance the possibilities of a successful pairing.

Additional Notes: Growth rate of spiderling *B. smithi* is slow compared to many species, but adulthood can be reached in five years (males) to seven years (females). This slow growth is characteristic of most long-lived species. In fact, female specimens which have been in captivity for over 25 years are common.

Greenbottle Blue Tarantula
Chromatopelma cyaneopubescens
Venezuela

Comments: *C. cyaneopubescens* is without a doubt one of the most beautiful species being kept in captivity.

Enclosure: *C. cyaneopubescens* in found in humid tropical forest areas but their enclosure in captivity can be a basic terrestrial setup. Substrate can be a relatively

Photo by Bill Korinek.

dry mix of sand and peat moss. A curved cork bark shelter should be added and the area underneath this bark should be sprayed once or twice each week. A live or artificial plant can add to the beauty of the Greenbottle Blue's enclosure. They produce a fair amount of webbing, but a keeper can gently remove the webbing on the plants to make sure they thrive in the enclosure.

Feeding: Greenbottle Blues feed on crickets and roaches.

Sexing: Mature male *C. cyaneopubescens* are strikingly colored in a more vibrant rendition of their normal coloration. They have an overall metallic look with longer, darker legs.

Breeding: Adult males should be carefully introduced into the female's enclosure after he has produced a sperm web. The male can be protected with a piece of cardboard or other tool if he is to be used for further breeding attempts. Once mating occurs, the female should be fed more heavily and a variety of prey items.

King Baboon Spider
Citharischius crawshayi
Kenya and Tanzania

Comments: At one time, the King Baboon Spider was the most sought after species for collectors. Their velvety terra cotta-colored hair and incredible display behaviors keep them popular among collectors. Unfortunately, like most baboon spiders, it is very aggressive and can be considered a potentially dangerous species. Adult female King Baboon Spiders can reach lengths of 7" to 8".

Enclosure: In nature, the King Baboon Spider is found in deep burrows associated with rock piles or the bases of trees. In captivity, they should be kept in a 15 to 20-gallon terrarium with a secure lid. This enclosure should have a deep substrate of at least 10" of compacted sand and peat moss, clay soil, or potting soil. They should be allowed to dig and maintain a burrow which

will help meet their need for a warm, humid retreat. Warmth in the 78° to 82° F range and a humidity of 75 to 85% is best for this burrowing species. A shallow water dish should be added to the enclosure.

Feeding: There is a tendency by keepers to feed their large, aggressive species lots of live mice. We suggest feeding baboon spiders a diverse diet consisting of adult crickets, grasshoppers, roaches, *Tenebrio* larvae, and only occasional feedings (once a month or less) of mice. We feel that this variety more closely mirrors the diet of tarantulas in nature and will keep them healthier and well-conditioned for potential breeding.

Sexing: Mature males are smaller than females, with longer, furrier legs and a bright red-orange coloration. *C. crawshayi* males have no tibial hooks to aid in breeding.

Breeding: Adult males should be carefully introduced into the female's enclosure after he has produced a sperm web. The males will be quickly attacked if the female is not interested in mating. He may be savagely killed and eaten unless a keeper can protect him with a piece of cardboard or other tool if he is to be used for further breeding attempts. Once mating occurs, the female should be fed a variety of prey on a more frequent schedule (up to three times a week). An egg sac will be produced from a month to eight weeks after mating. The sac, typical for baboon spiders, is suspended in a web hammock and will incubate for five to eight weeks. The babies emerge and molt and will begin spreading throughout the enclosure. When this occurs, they should be separated into individual containers as they will begin cannibalizing each other soon. Young baboon spiders can be easily raised in plastic vials or jars with a deep substrate (1/3 to 1/2 of the height of the enclosure) and a small clump of sphagnum moss or piece of leaf on top to hold some humidity in their environment.

Additional Notes: Unlike most baboon spiders, *C. crawshayi* is very difficult to inspire to breed, even when a healthy well-conditioned adult pair is available. The females are extremely aggressive to any males (and anything else) that come close to their burrow. Once an egg sac is produced, the female is reportedly a good parent and the process moves forward without problems. The spiderlings feed well and grow rapidly. Despite its incredibly aggressive nature, the King Baboon Spider has become a staple in the hobby and captive-produced spiderlings are occasionally available.

Rose-haired Tarantula
Grammostola cala
Chile

Comments: The Rose-haired Tarantula has been an important spider in our hobby for more than thirty years. As one of the most docile and hardy spiders being kept, it has been a standard in pet shops and science classrooms across the planet.

Enclosure: The Rose-haired Tarantula thrives in a simple enclosure as long as its need for temperatures in the 78° to 82° F range and they should have access to a constant supply of water. Though not necessary, we suggest adding a cork bark shelter, and other interesting decorations to the enclosure.

Feeding: Rose-haired Tarantulas feed well on a variety of insect prey including crickets, grasshoppers, locusts, roaches, and others.

Sexing: Mature male *G. cala* are much more colorful with an overall metallic rose to pink coloration. Their legs are much longer, thinner, and more hairy.

Photo by Bill Korinek.

Breeding: The Chilean Rose-hair can be bred in captivity if certain pre-mating conditions are maintained. The females seem to benefit from a cooling period of a couple of months prior to mating. Once a mature male is produced, and he makes a sperm web, he should be introduced into the female's enclosure. He will approach the female's shelter cautiously, tapping and vibrating his legs. The female will be "lured" out of her burrow or shelter and the male will typically lunge forward to use his hooks to hold the female's chelicerae and to push her into an almost upright position to give himself access to the female's epigyne for mating. A mature male can be introduced to multiple females or can be reintroduced to a female to enhance the possibilities of a successful pairing. Typically, the male will die in the weeks following a successful mating.

Colombian Giant Tarantula
Megahobema robustum
Colombia

Comments: *M. robustum* is a large and stocky member of this genus. A freshly molted adult specimen is very impressive and is the highlight of several collections in the United States. Adult specimens are a beautiful maroon to dark brown with long, reddish orange legs.

Photo by Bill Korinek.

Enclosure: This spider does better in a humid vivarium, but good ventilation is also required. Proper conditions are best accomplished with occasional mistings with warm water rather than the use of a damp substrate which will promote the growth of molds. Live plants can be added to enhance its enclosure and to aid in the retention of moisture.

This spider enjoys digging in the soil of its enclosure and it will only occasionally make use of a shelter. We suggest a temperature range of 75° to 80° F. If a large, flat water dish is always available, larger specimens of *M. robustum* will be forgiving of a lack of high ambient humidity.

Feeding: *M. robustum* will feed on crickets, grasshoppers, locusts, and small mice.

Sexing: Males have long fuzzy legs and are more strikingly colored than adult females.

Breeding: Tapping and vibrating is exchanged and the female will usually rush up to the male. If she is receptive, mating takes place without incident. If she is unreceptive, she will usually not signal by tapping and may quickly kill the male unless a keeper is diligent. Successful matings typically lead to egg sacs in 10 to 12 weeks. An egg sac can be expected to contain 100 to 150 eggs and to incubate for five to six weeks.

Additional Notes: Unfortunately, the rarity of this spider is also due to the fact that the spiderlings have proven very difficult to raise. Attempts worldwide have led to extremely high mortality in spiderlings and juvenile specimens. This high mortality is most definitely due to problems in humidity. Apparently, the young spiders have a fine margin of acceptable limits. We have had success raising *M. robustum* spiderlings using taller containers such as pint plastic jars. A deep soil substrate was added and a dime-sized screen-covered hole was placed in the lid. The added air space, moist substrate, and ventilation seemed to greatly aid the survival and growth of the spiderlings.

Blue Bloom Tarantula
Pamphobeteus nigricolor
Colombia, Ecuador, Peru, and Bolivia

Comments: *P. nigricolor* is a striking species. It is quite nervous and aggressive and will readily flick urticating hairs and will bite. These large species should be considered display animals rather than handleable pets.

Enclosure: As large forest spiders, *Pamphobeteus nigricolor* should be offered a large enclosure with plenty of extra floor space. The enclosure should be kept warm (78° to 82° F) and the substrate should consist of a thick (4") layer of lightly damp-ened sand and peat moss with a covering of sphagnum moss, cypress mulch, or leaves. A complex enclosure is not necessary, but this species will need a section of curved cork bark or other shelter, especially during molting. Misting with water once a week is recommended and a small water dish should be kept in the enclosure.

Photo by Bill Korinek.

Feeding:
This species should be offered a variety of larger prey once or twice each week. They will feed on adult crickets, grasshoppers, roaches, and will even eat lizards and small mice.

Sexing: The males of *Pamphobeteus* species are without a doubt some of the most beautiful spiders in the world. Mature *P. nigricolor* are no exception. On its final molt, the male appears as a velvety black spider with an incredible metallic violet starburst on the carapace and a eye-popping metallic violet sheen on the legs.

Breeding: A mature male, once he has successfully completed a sperm web, should be carefully introduced into the female's enclosure. The male can be protected with a piece of cardboard or other tool if he is to be used for further breeding attempts. One should be careful not to distress or "anger" a female while manipulating the male or to cause the male to race over to the female while escaping your prodding. Calm and quiet is called for. Once mating occurs, the keeper should continue feeding the female a variety of prey items in hopes of receiving a large egg sac full of fertile eggs.

Indian Ornamental Tarantula
Poecilotheria regalis
India

Comments:
It might be argued that *P. regalis* (along with *B. smithi*) is now the most well-known species of tarantula in the hobby.

Enclosure:
The Ornamental Tarantulas should be kept in a large, vertically oriented enclosure. A modified aquarium or large plastic tub with plenty of ventilation works best. If the enclosure becomes too dry, the spiders will not do well. We recommend that a keeper maintains 80° to 85° F with a humidity level of 75 to 85% for this species. One way to overcome the dilemma of high humidity and high ventilation is to use a deep (4 to 5") substrate of damp sand and peat moss and provide several live plants within the enclosure. These plants can be placed within the enclosure still in their pots or can be planted in the deep substrate.

Photo by Bill Korinek.

We suggest adding one or two shallow water dishes and misting the entire enclosure once a day to every other day. The cage should be allowed to dry out in-between mistings.

Feeding: *Poecilotheria* species feed eagerly on crickets, grasshoppers, and roaches.

Sexing: Mature males are slender and long-legged compared to females and they have no tibial hooks for mating.

Breeding: Adult males should be carefully introduced into the female's enclosure after he has produced a sperm web. The female should be very well-fed before any introductions as *Poecilotheria* females are notorious for attacking and eating males before any mating can occur. If both male and female are well-fed, success is more likely. The male will typically slowly and cautiously approach the female and will tap his front legs vigorously to announce his arrival as a potential mate and not a prey item. The female will typically signal back, but not as vigorously as the male. If the female does not respond with any tapping or signaling, she is very likely to attack and kill the male. Interestingly, males will often not mate within a female's enclosure, prefering to "lure" a female out of the enclosure. A keeper can line the outside of the female's enclosure with large pieces of driftwood and bark to form an arena in which the pairing will take place. In fact, recent successful breedings of *P. subfusca* have occured in this setup with the actual pairing taking place up and out of the enclosure (Korinek & Effenheim, pers. com.). Perhaps this space that is "neutral" for each member of the breeding pair keeps the female off guard and her instinctual habit of attacking anything that moves within the enclosure is somehow muted outside of her usual surroundings. The male can be protected with a piece of cardboard or other tool if he is to be used for further breeding attempts. Once mating occurs, the female should be fed in anticipation of an egg sac, which may contain 100 to 150 spiderlings.

Additional Notes: Though its bite can be medically significant and it is fast and somewhat aggressive, *P. regalis* is without a doubt one of the most exciting species being kept in captivity today.

Sun Tiger Tarantula
Psalmopoeus irminia
Venezuela, Guyana, and Northern Brazil

Comments: The Sun Tiger is a velvety black spider that is beautifully marked in orange stripes and yellow-orange leg markings.

Enclosure: The Sun Tiger Tarantula should be kept in a large, vertically oriented enclosure. A modified aquarium or large, well-ventilated plastic tub works best. We recommend that a keeper maintains 78° to 82° F with a humidity level of 75 to 85% for this species. A deep (4 to 5") substrate of damp sand and peat moss and several live plants within the enclosure should be offered. Add a shallow water dish and mist the entire enclosure once a day to every other day, depending on the outside conditions of the room in which the enclosure is located. The cage should be allowed to dry out in-between mistings.

Photo by Bill Korinek.

Feeding: *P. irminia* feeds aggressively on crickets, grasshoppers, and roaches.

Sexing: Male *P. irminia* (right) are darker and more slender and their markings are much more pale than those of the female.

Breeding: Adult males should be carefully introduced into the female's enclosure after he has produced a sperm web. Once mating occurs, the female should be fed more heavily and a variety of prey items.

Additional Notes: *P. irminia* is a fast and aggressive arboreal species. It will not hesitate to bite an unwary or careless keeper. Its bite could prove to be medically significant for some people.

Mombasa Golden Starburst Baboon Spider
Pterinochilus murinus
Zaire, Kenya, and Tanzania

Comments: The Starburst Baboon Spider is a yellowish tan to golden orange spider with clean, dark markings. In the Mombasa Golden morph, the black markings include the characteristic starburst pattern on the cephalothorax and dark chevrons on a golden orange abdomen.

Enclosure: *P. murinus* lives in heavily webbed burrows. It can live in a simple setup with 1-2" of substrate and with the addition of a piece of cork bark or artificial plant. It will take an existing shelter and build an extensive web system in which to live and to capture prey.

Feeding: Prey in nature includes insects, lizards, mice, and other small animals. In captivity, they do well on a diet of crickets and roaches.

Sexing: *P. murinus*, like most baboon spiders, mature at a surprisingly small size compared to breedable females. The small males are, however, eager and adept breeders.

Photo by Russ Gurley.

Breeding: Adult males should be carefully introduced into the female's enclosure after he has produced a sperm web. The male can be protected with a piece of cardboard or other tool if he is to be used for further breeding attempts. Once mating occurs, the female should be fed more heavily and a variety of prey items. As with most *Pterinochilus* species, mature males molt out quite small compared to females. This size difference ultimately leads to their demise post-mating. *P. murinus* has proven very easy to breed in captivity and females tend to be very protective of their egg sacs and young. Egg sacs commonly contain between 75 and 100 spiderlings.

Additional Notes: The bite of *P. murinus* has been well-documented and has produced a variety of symptoms in different people. These symptoms range from two to three weeks of extreme pain to loss of feeling in the bitten extremity for up to three months.

64   **Tarantulas and Scorpions**

Photo by Russ Gurley.

Goliath Bird-eating Spider
Theraphosa blondi
Suriname, Brazil, Guyana, and Venezuela

Comments: *T. blondi*, the Goliath Bird-eating Spider, along with its close relative, *T. apophysis*, are the world's largest species of spider. Its impressive size (8-9") makes it a popular species in collections. Unfortunately, it is rarely bred and most specimens that are available are imported from the wild.

Enclosure: As a very large species, *T. blondi* should be kept in a large enclosure. We suggest at least a 30-gallon terrarium or the largest plastic blanket boxes. The substrate should be deep (6-8") and can be peat moss and sand with a covering of cypress mulch. A large shelter should be offered in the form of a cork bark "cave" or half-burrowed clay pot. Though they are found in humid tropical forest areas, we suggest keepers maintain them on the dry side and spray them once or twice a week. A large diameter flat dish with fresh water should be available at all times.

Feeding: *T. blondi* should be fed a diverse diet consisting of

Tarantulas and Scorpions   **65**

adult crickets, grasshoppers, *Tenebrio* larvae, and only occasional feedings (once or twice a month) of small mice. We feel that this variety more closely mirrors the diet of tarantulas in nature and will keep them healthy.

Sexing: Mature males are slender and long-legged compared to females and they have no tibial hooks for mating.

Breeding: We feel that for successful reproduction, *T. blondi* females should be established in a large terrarium with at least 6-8" of substrate. A burrow can be started for them and the females will quickly finish this work and establish a burrow in the enclosure. This deep, secure burrow is an important step in successfully breeding this species.

Adult males should be carefully introduced into the female's enclosure after he has produced a sperm web. Usually, the male will approach the female's shelter and will tap the surface of the substrate to announce his arrival. In nature, this tapping no doubt lures the receptive female from her deep burrow. The female approaches the male and if she is receptive,will raise her front legs to give the male access for pairing. The male can be protected with a piece of cardboard or other tool if he is to be used for further breeding attempts. Once mating occurs, the female should be fed more heavily and a variety of prey items.

SUGGESTED READING

TARANTULAS

Baxter, R. 1993. Keeping & Breeding Tarantulas. Chudleigh publishing. Ilford, Essex, UK.

de Vosjoli, P. 1991. Arachnomania: Guide to Keeping Tarantulas and Scorpions in Captivity, Advanced Vivarium Systems. Mission Viejo, CA.

Gurley, R. 1993. Color Guide to Tarantulas of the World I. Living Art publishing. Ada, OK.

Gurley, R. 1995. Color Guide to Tarantulas of the World II. Living Art publishing. Ada, OK.

Hancock, K. and J. 1992. Tarantulas: Keeping and Breeding Arachnids in Captivity. R & A Publishing. Somerset, UK.

Klaas, P. 2001. Tarantulas in the Vivarium: Habits, Husbandry, and Breeding. Krieger Publishing. Melbourne, FL.

Marshall, S. D. 1996. Tarantulas and Other Arachnids. Barron's Educational Series. Hauppage, NJ.

Peters, H-J. 2001. Tarantulas of the World – Pocket Photo Identification Guide. Wegberg, Germany.

Reichling, S. 2003. Tarantulas of Belize. Krieger Publishing. Melbourne, FL.

Schultz, M. J. and S. A. Schultz. 1998. The Tarantula Keeper's Guide. Barron's Educational Series. Hauppage, NJ.

Smith, A. 1986. The Tarantula: Identification and Classification Guide. Fitzgerald Publishing. London, UK.

Smith, A. 1990. Baboon Spiders: Tarantulas of Africa and the Middle East. Fitzgerald Publishing. London, UK.

Smith, A. 1994. Tarantula Spiders: Tarantulas of the U.S.A. and Mexico. Fitzgerald Publishing. London, UK.

SCORPIONS

INTRODUCTION

Heterometrus spinifer, the Asian Forest Scorpion, is more aggressive and erratic than its similar looking kin, the Emperor Scorpion of Africa. Photo by Russ Gurley.

When the first of the terrestrial scorpions crawled out of a dark sea around 450 million years ago, it appeared much the same as today's scorpions – long, jointed legs, pedipalps with claws, and a long, articulated "tail" with a sting at the end. Most species of fossil scorpions that have been discovered vary only slightly from modern forms except that several are in excess of three feet long!

Today, there are roughly 1,500 recognized species of scorpions in the world of which some 25 species are dangerous to man. Worldwide, an estimated 1,000 people a year die from the stings of scorpions. Although these living fossils are considered primitive, they show some amazingly advanced adaptations including elaborate courtship behavior and maternal care of young.

Scorpion eyes are complex and unique. It is felt by some biologists that scorpions can even orient their activities at night by starlight. They can remain below freezing for weeks, they can withstand heat that will kill most living things, and some can be submerged under water for days and revive with little or no complications.

Scorpions are found on every continent except for Antarctica, having been introduced into New Zealand and England by man. They are found in deserts, in temperate forests, on mountaintops, along shorelines, and in tropical rainforests.

Today, as interest in the keeping of invertebrates grows, the scorpion has taken its place as a source of pleasure, curiosity, and fascination. Their ease of maintenance and the excitement that they can evoke will no doubt keep them popular with invertebrate keepers worldwide.

ANATOMY

The scorpion body is divided into a cephalothorax and mesosoma, or body, with the body thinning into a tail-like structure ending in a telson, where the sting is located.

On the cephalothorax are two large median eyes that are surrounded by two to six lateral eyes, depending on the species.

The mouth of the scorpion consists of a pair of chelicerae and a large pair of pedipalps. The chelicerae are made up of a movable "finger" and an immovable "finger" that form a sort of chewing pincer. The

The mouthparts, or chelicerae, of this drinking Emperor Scorpion are clearly visible. Photo by Russ Gurley.

pedipalps are leg-like and end in a pair of pincers which vary from long and slender to short and powerful. Legs of scorpions are similar to those of the spiders and other arthropods, having a coxa, a trochanter, a long femur, a tibia, and three tarsi.

The color of the body, pedipalps, and legs are of secondary importance in the identification of scorpions, as they often adapt to the color of their surrounding soils, sand, and wood, and also often vary to some degree with age.

A great deal of the scorpions show sexual dimorphism, or differences in the sexes. Males of a species often will have thinner pincers, and longer, narrower metasoma while the females tend to be stockier with shorter pincers and shorter metasoma. Also, there are occasionally differences in granulations and surface texture on the cuticle, or skin, of the scorpions that may vary according to sex, age, and the habitats where the individuals are found.

MOLTING

From six to ten times during the life of a scorpion it is required to

This *Opisthacanthus elatus* from Venezuela is in the middle of extricating itself from its molt. Photo by Eric Ythier.

molt its old cuticle in order to grow.

A scorpion, when ready to molt and shed its old exoskeleton, will seek out a dark and protected place. In the wild, this is in the burrow, under a board, stone, or cactus pad, in a clump of bromeliads, and in captivity it is generally under a shelter such as a piece of bark, clay pot, or stack of shale. To begin the molting process, the scorpion arches its body plates and moves slowly and rhythmically back and forth until the front and side portions of its old cephalothorax skin begins to split. The pedipalps and chelicerae are removed from the split in the old skin by a pumping action and slowly the cephalothorax, abdomen, and legs, one by one, are pulled from the molt.

The scorpion remains in its protective place for another day or two as the new soft skin hardens into the protective covering of the cuticle.

THE STING

The stinging apparatus of the scorpion is relatively simple. Inside the telson, located at the end of the metasoma, are two venom sacs. These two sacs are divided down the midline by a septum and each is surrounded by a double layer of cells. One layer is the granular secretory epithelium which secretes the venom and the other is a layer of connective tissue. A compressor muscle layer surrounds these two layers and presses them against the cuticle of the telson to extrude the venom by compressive forces when delivering a sting.

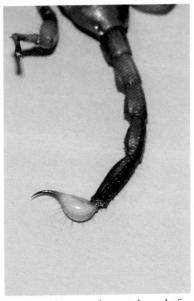

The sharp tip, or aculeus, at the end of the telson, is used to deliver the venom. Photo by Russ Gurley.

SCORPIONS of MEDICAL IMPORTANCE

Androctonus amoreuxi	dangerous
Androctonus australis	very dangerous
Androctonus bicolor	dangerous
Androctonus crassicauda	dangerous
Androctonus mauritanicus	dangerous
Buthus occitanus	dangerous
Centruroides exilicauda	moderately dangerous
Centruroides gracilis	dangerous
* Central America and Cuban specimens	
Centruroides limbatus	moderately dangerous
Hottentotta trilineatus	moderately dangerous
Leiurus quinquestriatus	very dangerous
Mesobuthus martensii	moderately dangerous
Parabuthus liosoma	moderately dangerous
Parabuthus granulatus	moderately dangerous
Parabuthus transvaalicus	moderately dangerous
Parabuthus mossambicensis	moderately dangerous
Tityus serrulatus	dangerous

Bücherl, W. and E. E. Buckley. 1971. Venomous Animals and Their Venoms. Academic Press.

Junghanss, T. and M. Bodio. 1996. Notfall-Handbuch Gifttiere. Georg Thime Verlag.

Keegan, H. L. 1980. Scorpions of Medical Importance. Fitzgerald Publishing.

Polis, G. 1990. Biology of Scorpions. Stanford University Press.

VENOM

The venom of scorpions varies considerably from species to species. There are, of course, dangerously venomous species whose venom rivals the world's most venomous snakes. Interestingly, venom potency can differ from population to population, even for the same species (Leeming, 2003).

The general rule is this: If a scorpion has thin, delicate-appearing pincers, then its venom is likely quite strong as it no doubt uses its venom to debilitate prey. If a scorpion has large, crushing pincers, then it is likely to use these to crush and disable prey and thus will have weaker venom.

Interestingly, in several species (*Hadogenes*, etc.) the male will sting a female during courtship to "anesthetize" or calm her down.

Scorpions are immune to their own venom unless they are stung in their basal ganglion on the underside of the scorpion's body. A species that regularly feed on other scorpion species (*Hadrurus*, etc.) will grab its prey, raise it up, and deliver a fatal sting into this area.

THE PECTINES

The pectines are comb-like sensory structures unique to scorpions. Over the years, many theories have been formulated as to the function of the pectines. Due to their look, many biologists originally believed that they were related to respiration. It is now known that the pectines are in fact sensory organs for detecting hormones, vibrations, and for sensing the temperature and humidity of the substrate (Leeming, 2003).

The pectines can be used in visually sexing many species of

The pectines are clearly visible on the ventral surface of this Blood Red Scorpion, *Babycurus jacksoni*. Photo by Russ Gurley.

Tarantulas and Scorpions

scorpions. Males tend to have more and longer "teeth", or setae, on their pectines and females tend to have fewer, much shorter setae on their pectines. These differences can be seen more easily by placing a specimen in a clear plastic shoebox or deli cup and looking at the ventral surface.

THE INTEGUMENT

The hard exoskeletal cuticle of the scorpion has some very important functions. The first of these functions is simply to provide protection from attack by predators. Scorpions are fed upon by many creatures in their natural habitats such as birds, mammals, and reptiles. The hard, outer cuticle helps to discourage some would be predators. The cuticle also acts as a barrier against water evaporation. This is important for desert-dwelling species, which can dry out very quickly in the heat of their environment.

FLUORESCENCE

Most people know that scorpions glow quite intensely and quite eerily under a UV-emitting light, from the beautiful ghostly blue of

The physiology of luminescence under a UV light, as seen in this Tricolor Scorpion, *Opistophthalmus wahlbergii*, remains one of the mysteries of the scorpion world. Photo by Russ Gurley.

Tarantulas and Scorpions   75

the Emperor Scorpion, *Pandinus imperator* to the fluorescent green of many desert species. Many researchers locate scorpions, even those many feet away with a handheld battery-operated UV bulb. An enclosure full of Emperor Scorpions, illuminated under a UV-emitting bulb in a dark room at night can be an exciting and somewhat disturbing "night light".

Though the actual system that allows a scorpion to fluoresce remains a mystery, recent studies show that the fluorescing is involved with the outer layer of the exoskeleton. A freshly molted scorpion does not fluoresce until its outer layer hardens.

Chapter SIX: SCORPIONS
IN CAPTIVITY

Euscorpius sicanus, a species commonly found in the crevices of stone walls and fences in Malta, is rare in captive collections. Photo by Jan Ove Rein.

Scorpions thrive in a variety of enclosures as long as their basic needs are met. We feel that glass terrariums and plastic tubs make excellent enclosures for keeping scorpions.

Glass terrariums can be excellent enclosures for scorpions. We suggest some of the modified, shorter tanks such as ½ 10-gallons, ½ 15-gallons, or 20-gallon long terrariums (12" x 12" x 20") for larger specimens. Typically terrariums are relatively inexpensive, available in a variety of sizes, and look nice when set up in a special part of a keeper's home. Secure screen tops are available for these glass terrariums and are usually easy to find at local pet stores. Plastic shoe boxes are also popular. They do not make exciting displays, but they are inexpensive and hold plenty of specimens in a small space. They are also easy to clean and sterilize.

This long terrarium provides a good home for Flat Rock Scorpions (or other rock-dwelling species). The bulb overhead provides heat which is so important in keeping this species healthy and feeding. Photo by Russ Gurley.

Substrate

Many keepers are choosing naturalistic setups for their invertebrate pets. Substrates are an important addition to the look of a naturalistic setup and seem to be important for captive scorpions. There are a variety of colored sands on the reptile market that can make beautiful naturalistic setups.

We suggest using a mixture of sand and peat moss. The peat moss will hold in some moisture. For those scorpions living in arid environments, we use a

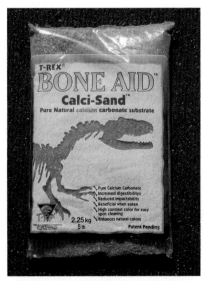

Calci-Sand® by T-Rex can be used to make an exciting naturalistic enclosure for species found in areas with colorful substrates such as Australia and parts of Africa.

A beautiful and creative enclosure for forest-dwelling species. Photo by Dan Read.

little more sand and for those living in humid forest or tropical areas, we add a little more peat moss to the mixture.

Cage Decorations

There is no doubt that the addition of plants, driftwood, cork bark, stable rock piles, and other cage decorations is important in keeping captive scorpions healthy and stimulated. These decorations can provide important hiding places, allowing the animal to feel secure and producing an area for safe molting. Cork bark shelters are ideal shelters for most species. Unfortunately, they are such excellent shelters that one rarely sees scorpions when they have bark in their enclosure. (Most keepers will use a fixture with a red bulb for viewing their nocturnal pet scorpions at night.) If you collect bark and other decorations from nature, be sure that they come from an area that is not sprayed with pesticides and that they are non-toxic. Rocks and stones can be sterilized in a light bleach solution (one part bleach to five parts water) and then scrubbed with a soap-filled sponge and rinsed thoroughly before they are added to the enclosure.

Be sure that any cage decorations are firmly resting on the bottom of the enclosure and that the substrate is pushed up against the object. If a scorpion digs under a flat rock or other heavy object, the object might settle onto the scorpion and crush it against the bottom of the enclosure.

Feeding

In nature, scorpions feed on a wide variety of invertebrates and the larger species will even take small reptiles and even small mammals. Some species are specialists, feeding on shoreline gastropods, other scorpions, centipedes, etc.

In captivity, most scorpions, such as this *Parabuthus* species, will feed on crickets, roaches, and other insect prey. Photo by Russ Gurley.

In captivity, most scorpions readily feed on the insects that are available to hobbyists through pet stores and commercial feeder suppliers. Crickets, roaches, mealworms, waxworms, and even pink mice are taken. We feel that offering the largest variety of prey items possible will keep your captive scorpions healthy.

We feel that a keeper should not feed their scorpions insects from the outdoors. Wild insects, especially crickets, are often plagued with internal parasites that may infect and kill your scorpion pets.

Chapter SEVEN: Breeding

Breeding in most scorpions begins with a "dance" involving grasping of pincers, locking of mouthparts, and gentle pushing and pulling. This is a pair of Striped Scorpions, *Centruroides vittatus*. Photo by Eric Ythier.

In scorpion species, the male initiates breeding. He slowly and carefully approaches the female's shelter. If the male does not approach with the proper signals, the female may retreat or chase the male away. Depending on the species, the male will "judder" (shake), tap his pincers on the shelter or ground, and/or bang his tail on the ground to announce his intentions of mating to the female.

If the female is receptive, she will either come out of her shelter, allow the male to drag her out of her shelter, or allow the male to enter her shelter. The male will grasp the female's pincers and in some cases, they will lock mouthparts. After a short "dance" of pushing and pulling, the male will drop a spermatophore, a packet of sperm, onto the substrate or an object in the enclosure such as a flat stone. The spermatophore is sticky at its base and has small hooks at its top. The male will carefully maneuver the female over the spermatophore and the hooks at its top will hook onto the female's genital opening. Once the female picks up the spermatophore, it breaks open and releases the sperm.

A female may produce several litters of young from sperm stored from a single pairing.

Sexing

Most species of scorpions exhibit sexual dimorphism, or differences in the body shapes of the males and females. In some species, the differences are very subtle and in others the differences are more distinct. In these species, typically males are longer and more slender and they may have much longer and thinner tails. The females tend to be more robust and have shorter, thicker tails.

Sometimes, male scorpions, such as this *Centruroides vittatus*, have elongated tails and a thinner build than females of the species. Photo by Jan Ove Rein.

Another indication of a scorpion's sex can be the characteristics of its pectines, the comb-like structures found on the undersides of their bodies. Typically, the "teeth" or setae of the male's pectines are longer than those of the female's.

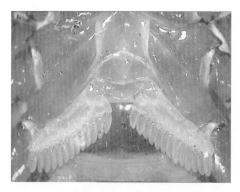

The "teeth" on the pectines of this *Tarsoporosus* species are longer and more numerous in the male (right). Photo by Eric Ythier.

Chapter EIGHT: Caring for Young Scorpions

This *Liocheles* species will carry her young on her back until their exoskeletons harden and they are capable of fending for themselves. Photo by Jan Ove Rein.

Scorpions give birth to live young.

Gestation in scorpions may last only a few weeks to as long as eighteen months. A scorpion may halt development of the embryos when conditions are harsh. When the time is right (usually when warm, rainy weather approaches), development continues. In most environments, the birth of baby scorpions will coincide with the seasons in which the insect populations will be flourishing. In captivity, captive production may be enhanced by following a dry period after mating with a more humid time and plenty of feeding during a warmer, more humid time.

When birth of the young is imminent, the female typically retreats to a safe shelter within its enclosure. Many species will even push substrate to cover the entrance to their shelter under flat stones or under pieces of cork bark. Within the safety of the shelter, the female will raise her body (known as "stilting") and bend the first pair of legs under the abdomen to form a basket.

This basket is used to catch the scorplings (babies) as they are born. Litters can be from eight to as many as fifty scorplings. They are immediately drawn to the mother's back by pheromones. Upon the mother's back, the babies receive protection. The female not only protects the scorplings from predators, but insures that they remain warm and humid until their exoskeletons harden and they are able to fend for themselves.

The deep humid substrate of this simple enclosure provides an ideal home for an Emperor Scorpion. Photo by Russ Gurley.

At a stage when they are hardened up and ready to explore and to catch their own prey, they will climb down from the mother's back, seek out their own shelters, and begin independent lives.

Caring for Young Scorpions

As the young scorpions begin to disperse, a keeper should probably remove them to individual enclosures or they may begin to cannibalize each other. Young scorpions should be kept slightly more humid than adults until their exoskeletons develop a stronger ability to prevent the loss of moisture. If the young scorpions are not kept humid, they will die quite quickly in a dry, arid container.

The small scorpion enclosure can be a small plastic tub, deli cup, or plastic shoe box. The substrate should be a mixture of damp sand and peat moss and several small pieces of bark or leaves can be added to provide a variety of shelters.

Small scorpions will typically feed on small crickets, small mealworms, and similarly sized prey. Many of the smallest species will

Small scorpions can be raised individually in deli tubs. Photo by Russ Gurley.

require day-old crickets or even wingless fruit flies for their first meals.

In some species, an adult female scorpion will kill a prey item and leave it on the ground. The young scorplings will climb down from her back and feed on the dead insect. If a keeper is having difficulty getting young scorpions to feed, he or she may have to smash a cricket and place it near the small scorpions. If the come into contact with some of the insect's bodily fluids, they may be inspired to feed.

As they grow, scorpions should be moved to larger enclosures. Conditions should begin to mirror the enclosures used for housing adults of the species. Conditions within these enclosures can become more dry as long as a keeper is careful to spray the inside of the enclosure once or twice a week for tropical species and less frequently for those species living in arid conditions. We recommend adding a shallow water dish in all scorpion enclosures.

SPECIES ACCOUNTS

Photo by Russ Gurley.

Deathstalker Scorpion
Androctonus amoreuxi
Egypt, Israel, and Jordan

Comments: *Androctonus amoreuxi*, along with its close relatives, *A. australis* and *A. bicolor*, vie for the title of the most venomous scorpions in the world. We feel that they should not be kept as pets for any reason. They are dangerously venomous and an accidental sting could prove life-threatening.

Enclosure: This species can be established in a horizontal profile enclosure with a substrate of sand and peat moss. They prefer a warm environment (90° to 95° F by day) and can tolerate extremes of temperature. A shelter in the form of a firmly supported flat stone or a piece of cork bark should be added. A keeper should spray the enclosure once a month. A humid enclosure will kill these scorpions. Multiple specimens can be kept together, but they should be added to an enclosure at the same time.

Feeding: These scorpions feed aggressively on a variety of insect prey including crickets, roaches, and even mealworms.

Sexing: In these species, typically males are longer and more slender and they may have much longer and thinner tails. The females tend to be more robust and have shorter, thicker tails.

Breeding: Breeding in this species is typical for most scorpions. A courtship dance involves grasping of pincers and chelicerae. This courtship is followed by deposition of the spermatophore by the male. Most *Androctonus* species will require a flat stone for placement of the spermatophore for successful reproduction. A female may produce one or two litters of young from sperm stored from a single pairing.

Additional Notes: This species is often imported by reptile dealers in the United States and Europe. They should be kept with extreme care and this author recommends keeping them similarly to venomous snakes in secure, locking enclosures. Long padded forceps can be used to rearrange their captive environment and to remove bits of uneaten food and waste.

Blood Red Scorpion
Babycurus jacksoni
Kenya

Comments: This social species, though moderately venomous, is not as aggressive and easily aggitated as the species already mentioned.

Their beautiful reddish coloration and ease of maintenance make them an exciting pecies for keepers.

Photo by Russ Gurley.

Enclosure: This forest-dwelling species can be

established in a horizontal profile enclosure with a substrate of and and peat moss. They prefer a warm environment and will require a humid shelter. This shelter can take the form of a firmly supported flat stone or a piece of bark. A keeper should spray under the shelter once a week to keep this area more humid.

Feeding: These scorpions feed aggressively on a variety of insect prey including crickets and roaches.

Sexing: As with many species, males of this species are more slender with elongated, thinner tails. Adult females are stockier and have thicker, shorter tails.

Breeding: This species has proven easy to breed and captive-born scorplings are often available. Breeding in this species is typical for most scorpions. A courtship dance involves grasping of pincers and chelicerae. This courtship is followed by deposition of the spermatophore by the male and the uptake of the spermatophore by the female. A female may produce up to three litters of young from sperm stored from a single pairing. Once they leave the female's back, the scorplings do well together for a month or two and then will begin cannibalizing each other, unless they are offered an abundance of small insect prey.

Central American Bark Scorpion
Centruroides margaritatus
Mexico, Guatemala, Ecuador, El Salvador, Costa Rica, Colombia, Ecuador, Nicaragua, Panama, Dominican Republic, Venezuela, Jamaica, and Cuba

Comments: *C. margaritatus* is a small, flat species found in moist forest areas of Florida in the United States. It is commonly encountered under the peeling bark of palms several feet off the ground. Central American Bark Scorpions are also occasionally found under stones, fallen logs, and piles of palm fronds.

Enclosure: This species can be established in a horizontal profile enclosure with a substrate of sand and peat moss. They prefer a warm, humid environment. A shelter in the form of a firmly supported flat stone or a piece of bark should be added. A keeper should spray under the enclosure once a week to keep this area more humid than the rest of the enclosure. They can be kept communally if they are established in an enclosure at the same time.

Photo by Russ Gurley.

Feeding: *C. margaritatus* do well on an insect diet which includes crickets and roaches.

Sexing: In these species, typically males are longer and more slender and they may have much longer and thinner tails. The females tend to be more robust and have shorter, thicker tails.

Breeding: Breeding in this species is typical for most scorpions. A courtship dance involves grasping of pincers and chelicerae. This courtship is followed by deposition of the spermatophore by the male and the uptake of the spermatophore by the female. A female may produce one or two litters of young from sperm stored from a single pairing.

Striped Scorpion
Centruroides vittatus
Mexico and United States (Arkansas, Colorado, Illinois, Louisiana, Missouri, Nebraska, New Mexico, Oklahoma, Texas)

Comments: This small scorpion is found throughout Oklahoma and Texas in the United States. It is often found living communally in large numbers under flat stones near ponds and in rock-strewn cattle pastures. They are small, with adults reaching only 1½" in total length.

Photo by Bill Korinek.

Enclosure:
This species can be established in a horizontal profile enclosure with a substrate of sand and peat moss. They prefer a warm environment (80° to 82° F). A shelter in the form of a firmly supported flat stone or a piece of bark should be added. A keeper should spray under the enclosure once a week to keep this area more humid than the rest of the enclosure.

Feeding: In nature, *C. vittatus* feeds on worms, spiders, crickets, and other small invertebrates. In captivity, they feed well on crickets, roaches, wax worms, mealworms, and other small insect prey.

Sexing: Males are somewhat slender with elongated metasomal segments of the tail. Females are shorter and often bulkier than males.

Breeding: Courtship in *C. vittatus* has been studied fairly extensively by invertebrate biologists and is fairly typical for *Centruroides* species. Courtship is initiated by the female upon an encounter with a male. The male quickly grasps the pincers of the female and begins a shaking action known as "juddering". Then, after a short shoving match, the male deposits a spermatophore onto the substrate and positions the female over the packet of sperm. The female lowers her abdomen and picks up the spermatophore into her genital opening. The two separate and often beat a hasty retreat in opposite directions. This species will readily breed in captivity and after a gestation period of eight months, females produce from fifteen to forty babies.

Additional Notes: This small species has a powerful sting that often produces intense pain that lasts for several hours. Other than this pain, however, there are rarely any other medical complications.

An interesting feature related to the native habitat of *C. vittatus* is its ability to remain alive during extended periods of below freezing temperatures. Studies show that species capable of living through these conditions have a protein-like substance that allows them to survive by "trapping" ice crystals in their gut.

Flat Rock Scorpion
Hadogenes troglodytes
Botswana, Mozambique, South Africa, and Zimbabwe

Comments: *Hadogenes troglodytes* is one of the largest, longest, and most unusual scorpions in the world.

Enclosure: This species is a lithophilic, or rock-dwelling spe-

Photo by Bill Love.

cies. Its long legs and compact body with stiff setae and strong claws help it to scurry quickly over rocks and stones in its rocky home. In nature, it is commonly found under stones in scraped out areas and wedged into cracks in crumbling stone outcrops where. They should be kept warm (84° to 88° F) for most of the year with a cooling period each evening and during a "Winter" (65° to 68° F for one to two months). They like it dry, but a misting of water once every other week is appreciated.

Feeding: In captivity, *H. troglodytes* will eat crickets and other insects, including wax worms and mealworms. Keeping them hot (84° to 88° F) for a portion of the day will ensure that they feed well.

Sexing: This species is sexually dimorphic, with males easily distinguished from females. Males show extreme elongation of the tail.

Breeding: Courtship and mating are fairly typical for most scorpions. The adult male initiates mating. He faces the female and grabs her pincers. He reaches under the female with his tail and interlocks his chelicerae with the females and a tug-of-war begins. While moving back and forth, the male deposits a spermatophore onto the surface and positions the female over the packet of sperm. She picks up the packet into her genital opening. The litter of 20 to 30 scorplings are born around 18 months later. They remain with the mother for up to two months, making short excursions from her back to search for food.

For the first few instars, small finicky *Hadogenes* can be assist-fed by smashing a cricket and placing it in front of them. Again, if they are not warm enough (80° to 86° F), they will be reluctant to feed.

Additional Notes: A strange habit is noted in the reproduction of *H. troglodytes*. This species, perhaps as a water conservation measure in its dry, arid habitat, does not produce a great deal of the moisture that is normally associated with the birth of baby scorpions. Thus, the "ease" of birth is seemingly compromised and the birth of the small scorplings can take as long as ten days!

Desert Hairy Scorpion
Hadrurus spadix
United States (Arizona, California, Nevada, and Utah) and Mexico

Comments: The Desert Hairy Scorpions are the largest scorpions in North America. They are commonly available to scorpion enthusiasts.

Enclosure: The Desert Hairy Scorpion is found in burrows in sandy scrub areas. In captivity, they are hardy and can be kept in simple enclosures

Photo by Eric Ythier.

with a sand substrate and a simple shelter. They require very little moisture, and a keeper can simply add a small dish of water to the enclosure for a day or two once a month. Temperatures in the 84° to 88° F range are ideal with very low humidity. Night cooling in the 60s F is recommended during the "Winter".

Feeding: This scorpion feeds on large insects such as locusts and solifugids and is reportedly a major predator of smaller species of scorpions such as *Vaejovis spinigerus*. In captivity, they feed well on crickets and roaches.

Sexing: In this species, sexes are very similar, but males are longer and more slender and females tend to be more robust.

Breeding: Unfortunately, even with an abundance of these scorpions in captivity, very few successful captive breedings have been reported. Females are notorious for eating their young.

This species probably requires a cooling period to inspire successful mating, healthy scorpling production, and long-term health in captivity.

Additional Notes: *H. spadix*, like many scorpions, fluoresces under ultraviolet light. This glowing gives away their position at night by collectors carrying portable black lights and adds to the large numbers found in captivity.

H. spadix is aggressive and will readily sting if given the opportunity. The sting is painful, but usually of little medical importance.

Tricolor Scorpion
Opistophthalmus wahlbergii
Africa

Comments:
The Tricolor Scorpion is active burrower. It has an exciting range of coloration and though it is nervous, makes a wonderful display animal.

Enclosure:
Tricolor Scorpions live in burrows in sandy areas. It should be given a deep, compacted, dry substrate (6-8") in which it can

Photo by Russ Gurley.

burrow. 75° to 85° F is the ideal temperature range.

Feeding: There is a tendency by keepers to feed their large, aggressive species lots of live mice. We suggest feeding even large scorpions a diverse diet consisting of adult crickets, grass-hoppers, *Tenebrio* larvae, and only occasional feedings (once or twice a month) of mice. We feel that this variety more closely mirrors the diet of scorpions in nature and will keep them healthy.

Sexing: Males of *Opistophthalmus* species have longer and thicker tails and more elongated pincers.

Breeding: If not overly obese, and if kept in a well-suited vivarium, *O. wahlbergii* will breed and produce offspring in captivity. After a gestation period of seven months, a litter ranging in size from 15 to 40 young scorpions is produced. The female scorpion will feed her young by killing an insect and leaving it on the floor of their enclosure. The scorplings will then descend from the mother's back and feed on the dead insect.

Additional Notes: This species, along with the other *Opistophthalmus* species, is known to stridulate (making a hissing sound) loudly when disturbed. The sound is made when the scorpion is rubbing its chelicerae together.

Emperor Scorpion
Pandinus imperator
Democratic Republic of Congo, Côte d'Ivoire, Ghana, Guinea, Nigeria, and Togo

Comments: The Emperor Scorpion is probably the most recognizable scorpion species (and possibly the most commonly kept invertebrate) in the world. As thousands have been im-ported from Togo and Ghana, Emperors have flooded the pet trade and have become very popular due to their impressive size (up to 8") and formidable appearance. They are a large, shiny black scorpion with massive pincers and a thick robust body and tail. Despite their gruesome look, the sting of the Emperor Scorpion is mild. We, however, do not recommend handling any

Photo by Bill Love.

scorpions.

Enclosure: Emperor Scorpions live in burrows in moist forest areas. A temperature range of 78° to 85° F is ideal. Substrate needs to be clean, damp peat moss. The addition of live plants will help keep the humidity level in the recommended 80 to 85% range. A water dish should be available at all times.

Feeding: There is a tendency by keepers to feed their large, aggressive species live mice. We suggest feeding even large scorpions a diverse diet consisting of adult crickets, grasshoppers, Tenebrio larvae, and only occasional feedings (once a month) of mice. We feel that this variety more closely mirrors the diet of scorpions in nature and will keep them healthy.

Sexing: Male Emperors are typically smaller and more slender than the larger, bulkier females. Also, the pectines of males have longer setae.

Breeding: If not overly obese, and if kept in a well-suited vivarium, *P. imperator* will often breed and produce offspring in captivity. After a gestation period of seven to ten months, a litter ranging in size from 15 to 40 young scorpions is produced. The female scorpion will feed her young by killing an insect and leaving it on the floor of their enclosure. The scorplings will then descend from the mother's back and feed on the dead insect. After a month or two, the scorplings will begin to venture out from the mother's back. If given a large enclosure with a proper burrow, the young scorpions will co-habitat with the female.

Additional Notes: Even with a multitude of captive breedings, very few young Emperor Scorpions reach adulthood. Many die from molting difficulties. Inability to exit their old skins from too dry conditions is proving to be the most commonly encountered problem. Large vivaria with a deep, damp substrate of sand and peat with multiple shelters and live plants have proven best for raising these large, but delicate, babies to maturity.

Israeli Gold Scorpion
Scorpio maurus palmatus
Algeria, Egypt, Libya, Mauritania, Morocco, Senegal, Tunisia, Iraq, Iran, Israel, Jordan, Kuwait, Lebanon, Qatar, Saudi Arabia, Syria, Turkey, and Yemen

Comments: The Israeli Gold Scorpion, *Scorpio maurus palmatus*, is a desert species with several interesting habits. It is unusual in that it is often active by day, especially during the rainy

Photo by Bill Love.

season. This is most likely due to the fact that a large number of diurnal insects flourish in its habitat during this time of year.

Enclosure: The Israeli Gold Scorpion lives in burrows dug with its large claws in the sandy soil, usually along the sides of a ravine or permanent sand dune. They move pebbles as well as large amounts of sand in the construction of these burrows. If kept in a vivarium with a relatively deep sand and peat moss substrate and with plants and broken clay pots or similar shelters, it can be kept communally. The enclosure should be large and shelters numerous when keeping multiple specimens together. *Scorpio* species have been bred occasionally over the last few years.

Feeding: Probably the most unusual aspect of the life of *S. m. palmatus* is its habit of feeding on isopods and beetles along the coastline of the Mediterranean Sea in Israel and Egypt. In fact, it has been discovered searching through seaweed in beach areas that are at times covered by seawater (Polis, 1990).

Sexing: In these species, typically males are more slender and the "fingers" of their pincers are short than those of females.

Breeding: Mating and birthing seasons are usually synchronized so that the female scorpions are giving birth at the same time. After birth, the young scorpions ride on the mother's back for a week to as long as a month before venturing out and establishing their own burrows.

It has been shown through laboratory investigations that the mother scorpion releases a chemical onto her back that signals to the baby scorpions that they are to group together there. In these experiments, baby scorpions that are placed on another female's back quickly dispersed. When the chemical is taken from the mother scorpion and placed on another scorpion's back – even a different species – the babies clamber aboard and ride this unwitting imposter.

Stripe-tailed Scorpion
Vaejovis spinigerus
United States (Arizona, California, and New Mexico) and Mexico

Comments: *V. spinigerus* is a small species from the southern United States and northern Mexico. It is found in hot, dry areas and is commonly seen under stones, boards, and pieces of Cholla and prickly pear cactus. In many areas, *V. spinigerus* can be found in large numbers around abandoned barns and sheds, under the shingles and old lumber. It is a feisty captive.

Photo by Bill Korinek.

Enclosure: This species can be established in a horizontal profile enclosure with a substrate of sand and peat moss. They prefer a warm environment (78° to 84° F) and can tolerate extremes of temperature. A shelter in the form of a firmly supported flat stone or a piece of cork bark should be added. A keeper should spray under the enclosure once a week to keep this area more humid than the rest of the enclosure.

Feeding: This smaller species will feed on crickets and small roach species.

Sexing: In these species, typically males are longer and more slender and have longer and thinner tails. The females tend to be more robust and have shorter, thicker tails.

Breeding: Breeding in this species is typical for most scorpions. A courtship dance involves grasping of pincers and chelicerae. This courtship is followed by deposition of the spermatophore by the male and the uptake of the spermatophore by the female. A female may produce one or two litters of young from sperm stored from a single pairing.

Additional Notes: The hardy Stripe-tailed Scorpion is a belligerent little scorpion and seems always ready to attack. The sting is quite painful and reports of extensive pain have been associated with this species.

SUGGESTED READING

Bawaskar, H. S. 1999. Scorpion Sting - Clinical Manifestations, Management and Literature. Poular Prakashan.

Bullington, S.W. 1996. Natural history and captive care of the flat rock scorpion *Hadogenes troglodytes*. Vivarium 7(5), pp.18-21.

de Vosjoli, P. 1991. Arachnomania: Guide to Keeping Tarantulas and Scorpions in Captivity, Advanced Vivarium Systems. Mission Viejo, CA.

Fet, V., W. D. Sissom, G. Lowe, and M. E. Braunwalder. 2000. Catalogue of the Scorpions of the World (1758-1998). The New York Entomological Society.

Fet, V. and P. Selden (eds.). 2001. SCORPIONS 2001 - In memoriam Gary A. Polis. British Arachnological Society.

Francke, O. and S. Jones. 1982. The life history of *Centruroides gracilis* (Scorpiones, Buthidae). J. Arachnol. 10: 223-239.

Francke, O. and S. Stockwell. 1987. Scorpions from Costa Rica. Texas Tech Univ. Press. Lubbock, TX.

Gaban, D. 1997. On: *Opistophthalmus glabrifrons* (Peters). Forum American Tarantula Society 6 (6), p. 196.

Hull-Williams, V. 1986. How to Keep Scorpions. Fitzgerald Publishing. London, UK.

Keegan, H. L. 1980. Scorpions of Medical Importance. Fitzgerald Publishing. London, UK.

Leeming, J. 2003. Scorpions of Southern Africa. Struik Publishers. South Africa.

Polis, G. 1990. Biology of Scorpions. Stanford University Press.

Tikader, B.K. and D. B. Bastawade. 1983. Fauna of India. Scorpionidae : Arachnida. Vol. III: Scorpions. Zoological Survey of India. Calcutta, India.